Seeds, Science, and Struggle

Food, Health, and the Environment
Series editor: Robert Gottlieb, Henry R. Luce Professor of Urban and
Environmental Policy, Occidental College

Seeds, Science, and Struggle

The Global Politics of Transgenic Crops

Abby Kinchy

The MIT Press
Cambridge, Massachusetts
London, England

MIT Press books may be purchased at special quantity discounts for business or sales promotional use. For information, please email special_sales@mitpress.mit.edu or write to Special Sales Department, The MIT Press, 55 Hayward Street, Cambridge, MA 02142.

This book was printed on recycled paper and set in Sabon by the MIT Press. Printed and bound in the United States of America.

100700600&
Library of Congress Cataloging-in-Publication Data

Kinchy, Abby J.
Seeds, science, and struggle : the global politics of transgenic crops / Abby Kinchy.
  p.   cm. — (Food, health, and the environment)
Includes bibliographical references and index.
ISBN 978-0-262-01781-7 (hardcover : alk. paper) — ISBN 978-0-262-51774-4 (pbk. : alk. paper)
1. Transgenic plants—Risk assessment. 2. Plant genetic engineering—Political aspects. 3. Plant genetic engineering—Risk assessment. I. Title.
SB123.57.K48   2012
632′.8—dc23
2011053204

10 9 8 7 6 5 4 3 2 1

For my parents

Contents

Series Foreword

I am pleased to present the eighth book in the Food, Health, and the Environment series. This series explores the global and local dimensions of food systems, and examines issues of access, justice, and environmental and community well-being. It includes books that focus on the way food is grown, processed, manufactured, distributed, sold, and consumed. Among the matters addressed are what foods are available to communities and individuals, how those foods are obtained, and what health and environmental factors are embedded in food-system choices and outcomes. The series not only concentrates on food security and well-being but also regional, state, national, and international policy decisions as well as economic and cultural forces. Food, Health, and the Environment books provide a window into the public debates, theoretical considerations, and multidisciplinary perspectives that have made food systems and their connections to health and environment important subjects of study.

Robert Gottlieb, Occidental College
Series editor

Acknowledgments

I want to express my gratitude for the time, information, and insight provided by each of the people who shared their knowledge with me while researching this book. I met so many kind, helpful, and inspiring people during my field research in Mexico and Canada that it is not possible to thank them all individually. A few individuals and organizations stand out, however. Cati Marielle of Grupo de Estudios Ambientales in Mexico City, Álvaro Salgado of Centro Nacional para Misiones Indígenas in Mexico City, Don and Debbie Kizlyk of the Saskatchewan Organic Directorate, and Brenda Frick of the Organic Agriculture Center of Canada were remarkably generous with their time and welcomed me warmly to their communities. I am also indebted to Arcelia González Merino, my skillful research assistant in Mexico City, who was an enormous help, and became a wonderful friend and colleague too. I would also like to thank the faculty and graduate students in the biotechnology study group in the sociology department at the Universidad Autónoma Metropolitana, Unidad Azcapotzalco, particularly Michelle Chauvet, who made me feel included in the research community there.

I have been fortunate to have several mentors throughout the process of researching and writing this book. Above all, I am grateful to Daniel Kleinman, my graduate adviser, mentor, collaborator, and friend. It would be impossible to express how much I have valued working with and learning from him over the past decade. He not only steered me toward questions about science and politics as well as taught me how to look for the answers; he also provided crucial emotional support and professional advice throughout the long process of bringing this book into being. My thanks go to Jack Kloppenburg, whose brilliant analysis of the politics of genetic resources motivated me to pursue this line of study, and whose guidance was so important to me as a graduate student. Jane Collins, Gay Seidman, and Eric Schatzberg, my other committee members, deserve great appreciation for their key insights, questions, advice, and encouragement. I also wish to thank David Hess, an amazing mentor and colleague, for always advocating for me and this project, even when I had doubts. It has been such a pleasure working with him. And I must thank Rachel Schurman, who offered excellent advice when I really needed it.

This project received generous financial support from the Social Science Research Council, in the form of an International Dissertation Research Fellowship. A National Science Foundation Dissertation Improvement Grant (No. 0525799) also supported my research. In addition, I received a variety of grants and fellowships provided through the University of Wisconsin at Madison, including the Raymond J. Penn Scholarship, the Wisconsin Distinguished Graduate Fellowship Award, and the Latin American, Caribbean, and Iberian Studies Program Field Research (Tinker-Nave) Grant. I am humbled and extremely grateful that each of these institutions was confident in the potential contributions of my study.

Some of the research presented in this book was previously published and is used here with permission. In particular, much of chapter 3 appeared in the journal Mobilization as "Epistemic Boomerang: Expert Policy Advice as Leverage in the Campaign against Transgenic Maize in Mexico." Parts of chapters 1 and 2 were published in Agriculture and Human Values, in an article titled "Anti-Genetic Engineering Activism and Scientized Politics in the Case of 'Contaminated' Mexican Maize." Finally, key elements of the argument presented in chapter 2 were first developed in collaboration with Daniel Kleinman and Robyn Autry in the process of writing the article "Against Free Markets, Against Science? Regulating the Socioeconomic Effects of Biotechnology" for the journal Rural Sociology. I thank my coauthors for their gracious permission to draw from this collaborative piece.

Many people helped to transform my manuscript into a book. I wish to thank Clay Morgan, Robert Gottlieb, and three anonymous reviewers for their thoughtful, constructive feedback. Working with Deborah Cantor-Adams and others at the MIT Press has been a pleasure. I also thank Sabrina Weiss for offering research assistance as I was revising the manuscript.

I want to express my appreciation to all my colleagues in the Department of Science and Technology Studies at Rensselaer Polytechnic Institute, who created a stimulating, supportive environment in which to pursue the ideas in this book. In particular, Kim Fortun, Nancy Campbell, and Ned Woodhouse offered advice and encouragement on this project. Dean Nieusma and Michael Mascarenhas became good friends as we simultaneously encountered the challenges of being both scholars and parents of small children. I couldn't ask for a better group of people with whom to work, think, and write.

I counted on the support of many friends, both old and new, in the process of writing this book. In particular, I credit the members of various writing groups for keeping me relatively sane, organized, and happy over the years. These include Kaelyn Stiles, Amy Lang, Aya Hirata Kimura, Danielle Berman, Kate McCoy, Andrea Voyer, Nicole Breazeale, Jay Burlingham, Jan Fernheimer, Ellen Esrock, Katya Haskins, Jennifer Burrell, and Linnda Caporael. Gwen Sharp, Jason Delborne, Shobita Parthasarathy, Beth Berman, and Virginia Eubanks have offered both friendship and intellectual stimulation over the years. Emily Eaton and her family, Steve and Kim Weiss, Karina País, and Devorah Wolf all made my field research infinitely more enjoyable, and I learned a lot from them, too.

Finally, I want to thank my parents, Sue and Jerry Kinchy, for teaching me where food comes from and always being my biggest supporters. My husband, Nathan Meltz, endured my long absences, made excellent mix tapes, filled my office with his beautiful prints of maize and canola, and read every chapter cheerfully. I owe him my deepest and most sincere gratitude. My son, Aldo, has been remarkably patient with me. I hope that all the joy and optimism that he has brought into my life also comes through on these pages.

Acronyms

AAFC	Agriculture and Agri-Food Canada
APHIS	Animal and Plant Health Inspection Service
CASIFOP	Centro de Análisis Social, Información y Formación Popular (Center for Social Analysis, Information, and Popular Training)
CBAC	Canadian Biotechnology Advisory Committee
CBAN	Canadian Biotechnology Action Network
CBD	(UN) Convention on Biological Diversity
CEC	(North American) Commission for Environmental Cooperation
CECCAM	Centro de Estudios para el Cambio en el Campo Mexicano (Center for Studies for Change in Rural Mexico)
CENAMI	Centro Nacional de Ayuda a Misiones Indígenas (National Center for Assistance to Indigenous Missions)
CFIA	Canadian Food Inspection Agency
CFS	Center for Food Safety
CIBIOGEM	Comisión Intersecretarial de Bioseguridad y Organismos Genéticamente Modificados (Interministerial Commission for Biosafety and Genetically Modified Organisms)
CNCA	Consejo Nacional para la Cultura y las Artes (National Council for Culture and the Arts)
CNI	Congreso Nacional Indígena (National Indigenous Congress)
EIS	Environmental impact statement
EPA	Environmental Protection Agency

ETC Group	Action Group on Erosion, Technology, and Concentration
EZLN	Ejército Zapatista de Liberación Nacional (Zapatista National Liberation Army)
FDA	Food and Drug Administration
GATT	General Agreement on Tariffs and Trade
GE	Genetically engineered
GEA	Grupo de Estudios Ambientales (Environmental Studies Group)
GM	Genetically modified
GRAIN	Genetic Resources Action International
GURT	Genetic use restriction technology
INE	Instituto Nacional de Ecología (National Ecology Institute)
JPAC	Joint Public Advisory Committee (of the CEC)
LBOGM	Ley de Bioseguridad de Organismos Genéticamente Modificados (Biosafety Law for Genetically Modified Organisms)
NAAEC	North American Agreement on Environmental Cooperation
NAFTA	North American Free Trade Agreement
NFU	National Farmers Union (Canada)
NGO	Nongovernmental organization
OAPF	Organic Agriculture Protection Fund
PBR	Plant breeders' rights
PCR	Polymerase chain reaction
PROFEPA	Procuraduría Federal de Protección al Ambiente (Federal Attorney for Environmental Protection)
PUBPAT	Public Patent Foundation
RAFI	Rural Advancement Foundation International
SAGARPA	Secretaría de Agricultura, Ganadería, Desarrollo Rural, Pesca y Alimentación (Ministry of Agriculture, Livestock, Rural Development, Fisheries, and Food)
SEMARNAT	Secretaría de Medio Ambiente y Recursos Naturales (Ministry of Environment and Natural Resources)

SOD	Saskatchewan Organic Directorate
SPS	(WTO Agreement on the Application of) Sanitary and Phytosanitary Measures
UCCS	Unión de Científicos Comprometidos con la Sociedad (Union of Socially Committed Scientists)
UNAM	Universidad Nacional Autónoma de México (National Autonomous University of Mexico)
UNESCO	UN Educational, Scientific, and Cultural Organization
UNORCA	Unión Nacional de Organizaciones Regionales Campesinas Autónomas (National Union of Regional Autonomous Campesino Organizations)
UNOSJO	Unión de Organizaciones de la Sierra Juárez de Oaxaca (Union of Organizations of the Sierra Juarez of Oaxaca)
USDA	US Department of Agriculture
WTO	World Trade Organization

1

Introduction: Genes Out of Place

Possibilities made available by the new technologies are profound ones. But they are not always questions of risk. It may happen, nevertheless, that because "risk" was the focus of discussion of rDNA research and development in the early years, it will continue to shape discussions on this topic for decades to come. I could see, for example, the application of moral rules of the following kind: unless the development of a new genetic configuration can be shown to involve substantial quantifiable "risk," the development will be sanctioned for speedy implementation. If that happens, all the shortcomings of risk assessment will return to us with a vengeance.

—Langdon Winner, *The Whale and the Reactor*, 1986

In a move that caused jaws to drop in the farm industry, Agriculture Secretary Tom Vilsack has invited activists and biotech critics to shape the agency's regulatory decision on a biotech product. If the precedent stands, it could permanently politicize a system that is supposed to be based on science.

—"Ag Department Uproots Science," *Wall Street Journal*, December 26, 2010

Alfalfa is not typically a big news story, even though it is among the top four crops produced in the United States. It is a perennial grass, mostly grown to make hay for cows—one of the major inputs to the dairy industry that remains largely invisible to consumers. But in 2010, as the Supreme Court prepared to rule on a case involving a type of alfalfa that had been genetically altered to tolerate herbicides, the low-profile plant began to grab headlines. The case revolved around an important question: What is the right way for the government to assess and regulate genetically engineered (GE, or *transgenic*) organisms that will be released into the environment?[1] The *Wall Street Journal* argued that GE alfalfa should be commercialized as long as there is no scientific proof that it is harmful, and that it was inappropriate to include stakeholders such as organic food activists in the decision-making process. To the editorial writers' alarm, the US agriculture secretary seemed to be ready to listen to "political" perspectives on GE alfalfa.

The *Wall Street Journal*'s assertion is not unusual. Many observers of this and other controversies over GE crops frame the issue as a struggle to maintain the purity of science against political interference. Both advocates and critics of GE crops repeatedly call for "science-based regulation," and criticize what they view as political manipulation of the regulatory process. But what does it mean to "politicize" regulatory decisions about GE crops? What exactly are the political issues at stake when GE crops are considered for deregulation? And why should they not be relevant to regulatory decisions? In this book, I consider these questions by turning the conventional argument about the regulation of GE crops on its head. Rather than critique the *politicization* of the regulatory system, I take a close look at the *scientization* of public debate about biotechnology.

By scientization, I mean the transformation of a social conflict into a debate, ostensibly separated from its social context, among scientific experts. In the epigraph to this chapter, Winner accurately predicted the scientization of biotechnology governance in the United States, where GE crops are typically deregulated unless there is strong evidence of "substantial quantifiable 'risk.'" In this context, science and scientists are frequently considered to be the best possible arbiters of technological controversies, because they are assumed to produce objective, value-neutral assessments that do not favor one social group over another.[2] As critics have long pointed out, however, and as the case studies examined in this book will illustrate, agricultural biotechnology has a wide range of cultural, economic, and ethical implications that are not easily reduced to scientific calculations of risk.

I contend that the elevation of risk assessment over questions about the social desirability of transgenic crops has led to regulatory decisions that exacerbate serious problems facing the global agrifood system. This book contrasts the "science-based" rules for GE crops with the complex social consequences of releasing them into the environment. The shortcomings of scientific risk assessment are particularly visible in cases of genetic "contamination"—the commingling of GE plants with their non-GE relatives, leading to the unwanted presence of transgenes in plants or foods where they were not intentionally placed.[3] Throughout the book, I refer to these wayward transgenes as "genes out of place." As I will show in the subsequent chapters, genes out of place have led to major conflicts over intellectual property, organic standards, genetic diversity, global trade, corporate concentration, and maintenance of food traditions—conflicts that potentially undermine risk assessment as the basis for governing biotechnology.

I focus on two of the most high-profile cases of contamination of the last decade—unapproved transgenic material in traditional varieties of Mexican maize and the failure to separate GE canola from the rest of the seed supply in Canada. In both cases, activists connected to a variety of struggles for social change, including movements for the environment, global justice, genetic resources, organic farming, and indigenous rights, have persistently highlighted the socially problematic nature of GE crops. The chapters to come focus on these activists—the "antibiotech movement"—as they interact with national and international institutions that insist on science-based decisions about GE crops.[4] In these episodes of contention, farmers and activists question the science used in regulatory decisions and seek to broaden the range of considerations relevant to the governance of GE crops. For example, small-scale maize farmers in Mexico argue that the integrity of their traditions and stability of their livelihoods should be primary concerns. In response, as seen in a wide range of recent debates, from tobacco to gas drilling, industry groups assert that decisions must be based on "sound science." These moments of conflict indicate that the scientization of governance, far from creating a neutral basis for decisions, has the effect of excluding less powerful actors from policy debate about the social organization of agriculture and appropriate role of GE crops.

The case of alfalfa in the United States offers an example of some of the issues that are at stake in these debates. In 2004, Monsanto Company, a manufacturer of agricultural chemicals and GE seeds, sought regulatory approval of a new type of alfalfa, called Roundup Ready, in the United States. This alfalfa was genetically engineered to be tolerant of the herbicide Roundup, also manufactured by Monsanto. Because of this tolerance, farmers could spray herbicides to kill weeds without damaging their alfalfa crop. The Animal and Plant Health Inspection Service (APHIS), the US regulatory agency responsible for determining which GE crops require regulation, decided that Roundup Ready alfalfa would have no significant impact and could be commercialized without further study.[5] Some producers of alfalfa and environmental advocacy groups strongly disagreed with that decision, though. In early 2006, a group of plaintiffs, including two alfalfa seed producers, the National Family Farm Coalition, the Sierra Club, the Center for Food Safety, and other environmental advocacy groups, brought a lawsuit against the US regulatory agencies responsible for governing GE crops. The plaintiffs alleged that APHIS violated US environmental law because it had not produced an environmental impact statement (EIS) before deregulating Roundup Ready alfalfa. They asked the court to stop the commercial cultivation of GE alfalfa until the US Department of Agriculture (USDA) completed a full environmental impact assessment, including an analysis of the socioeconomic impacts (Center for Food Safety 2006). Because organic dairy producers expect to buy "non-GE" alfalfa, unwanted contamination or "commingling" of GE and non-GE crops would likely have significant economic consequences for farmers—potentially threatening the organic dairy industry as a whole (Center for Food Safety 2011b). Additionally, the plaintiffs pointed to other new agricultural and environmental problems resulting from the heavy use of herbicide-tolerant GE crops, particularly an increase in the use of herbicides and the development of herbicide-resistant weeds (ibid.).

A California federal district court ruled that the USDA was required to carry out a more comprehensive EIS that would include a study of the impacts on the production of non-GE varieties of alfalfa, among other environmental impacts. The ruling put sales of Monsanto's alfalfa seeds on hold until APHIS had completed the necessary EIS in 2010. The final document, over two thousand pages long, considered options for growing GE alfalfa commercially without restrictions as well as with partial restrictions, all aimed at minimizing the contamination of non-GE fields (US Department of Agriculture 2010). On the release of the EIS in December 2010, and hoping to put a stop to future litigation, Secretary of Agriculture Vilsack proposed the creation of "coexistence" policies that would permit the production of both GE and non-GE alfalfa. The USDA convened a stakeholder meeting aimed at beginning a dialogue about the coexistence of different varieties of agriculture using GE and non-GE crops. Vilsack (2010) explained the approach he wished to take in an open letter:

As a regulatory agency, sound science and decisions based on this science are our priority, and science strongly supports the safety of GE alfalfa. But, agricultural issues are always complex and rarely lend themselves to simple solutions. . . . By continuing to bring stakeholders together in an attempt to find common ground where the balanced interests of all sides could be advanced, we at USDA are striving to lead an effort to forge a new paradigm based on coexistence and cooperation.

Both critics and advocates of GE alfalfa criticized the USDA's handling of the alfalfa issue. One of the advocacy organizations that pursued the litigation, the Center for Food Safety (2011a), pointed out flaws in the science presented in the EIS, saying that the assessment was still inadequate. Advocates of biotechnology, on the other hand, vigorously criticized the "bad faith, obstructionist" use of federal environmental law to unreasonably drag out the risk assessment process (Conko and Miller 2010). Supporters of GE alfalfa indicated that the weed problem was manageable and the economic effects of contamination were beyond the scope of regulatory authority. US representative Frank Lucas of Oklahoma, chair of the House Agriculture Committee, along with US senators Saxby Chambliss and Pat Roberts, of Georgia and Kansas, respectively, sent a letter to Vilsack criticizing the move to consider coexistence: "Decisions should be based on science with other factors more appropriately considered in the marketplace. Our government fought diligently to preserve the integrity of science-based decision making in the World Trade Organization and the success in that body should not be so casually set aside (House Committee on Agriculture 2011).[6]

Vilsack subsequently backed off from the idea of pursuing coexistence policies and said the USDA would authorize the unrestricted commercial cultivation of the controversial crop. As of late 2011, the status of GE alfalfa is still in flux, as critics continue to pursue litigation to halt its commercialization.

The GE alfalfa controversy highlights one of the major questions associated with releasing GE plants into the environment: Should the government create measures to preserve the cultivation of non-GE crops? If so, on what grounds, and at what cost to the biotechnology industry? Additional controversies about genes out of place revolve around related, but different questions. Who, if anyone, can be held responsible when transgenes contaminate crops meant to be non-GE? Do patents on transgenic material extend to the offspring of GE crops, even when they accidentally wind up on another farmer's land? What if transgenic material crosses national borders, into areas where GE crops are highly regulated or face strong cultural opposition? How these questions are answered, and whose opinions matter in reaching those answers, will have an enormous impact on the future of food.

What Are GE Crops?

To understand the controversies discussed in this book, it is necessary to know some basics about agricultural biotechnology. For thousands of years, humans have used breeding techniques to intentionally change the genetic properties of animals and plants. Such techniques include selection, crossbreeding, and hybridization as well as the more recent use of radiation or chemicals to induce random mutations that may turn out to be useful. In a broad sense, therefore, most domesticated plants and animals are "genetically modified." Yet the techniques used to produce GE organisms are novel and distinct from the breeding methods of the past.

In genetic engineering, scientists directly manipulate an organism's DNA, the genetic information within every cell that allows living things to function, grow, and reproduce. Segments of DNA that are known to produce a certain trait or function are commonly called "genes." Genetic engineering techniques enable scientists to move genes from one species to another. After isolating and removing a gene from a cell of one organism, the gene can be transferred to another organism, using biological vectors such as plasmids (parts of bacteria) and viruses to carry foreign genes into cells, injecting genetic material containing the new gene into the recipient cell with a fine-tipped glass needle, using chemicals or an electric current to create pores in the cell membrane that allow entry of new genes, or a "gene gun," which shoots microscopic metal particles, coated with genes, into a cell (Union of Concerned Scientists 2003). Moving genes from one species to another results in novel genetic combinations, giving the recipient organism characteristics associated with the newly introduced gene.

Ever since scientists first developed genetic engineering techniques in the 1970s, there has been enormous hope and an expectation that they would bring useful improvements to agriculture. Genetic engineering allows plant breeders to create plants with characteristics that would have been difficult or impossible to develop through traditional breeding. Consistent with some of those expectations, genetic engineering has been used successfully to prevent certain plant diseases (such as the papaya ring spot virus), resist insect pests, and facilitate weed control. Genetic engineering could potentially increase nutritional content and drought resistance as well as address other agronomic challenges (Wu and Butz 2004). Frequently, advocates of GE crops argue that biotechnology is necessary in order to meet the food needs of a growing population—a claim that is vigorously debated.[7] Even if all the hype about the future of GE crops proves true, the reality today is that the vast majority of the GE crops being grown are soybeans, corn, and cotton, primarily featuring two transgenic traits: herbicide tolerance, and resistance to insect pests. Together these three GE crops totaled 95 percent of the global GE crop area in 2010 (James 2010b).[8] From 1994, the year that the first GE plant reached the market (Calgene's Flavr Savr tomato, a commercial failure), to 2010, the cultivation of GE crops has grown rapidly. In 2010, 148 million hectares of GE crops were grown in twenty-nine countries (ibid.). Most GE crops, however, are grown in just a few countries. The United

States, Brazil, and Argentina combined cultivate 77 percent of the land dedicated to GE crops, with the United States far in the lead with over 66 million hectares.

The lack of diversity in the GE crops that are being commercially produced is mirrored in the lack of diversity in the seed industry. Monsanto and DuPont, both originally in the agrichemical business, are now the two largest global seed companies, controlling the development of engineered traits, incorporation of those traits into existing germplasm (seeds), and provision of chemicals to complete the "package" (Boyd 2003). By 2009, just four companies controlled 50 percent of the world's market in seeds that are subject to intellectual property protection—and 43 percent of the entire world seed market (Hubbard and Farmer to Farmer Campaign on Genetic Engineering 2009). The concentration is more extreme in certain crops, such as corn, where one company, Monsanto, controls about 60 percent of the seed market in the United States (ibid.).[9]

The development trajectory of GE crops has been consistent with industrialization processes in agriculture, through which elements of agricultural production (such as nutrient cycles, pest control, irrigation, energy, and motive power) are transformed into standardized industrial activities (Goodman, Sorj, and Wilkinson 1987). The dominant GE crops are designed to be most beneficial in large-scale monocultures that rely on mechanization, chemical inputs, and elimination of pests. They also provide a way for seed producers to extract profits from genetic resources by stopping farmers from using farm-saved seeds. Historically, farmers have saved and reproduced seeds from their own harvests, reducing or eliminating the need to purchase seeds. For decades, for-profit seed companies have tried to turn seeds into an industrial input that farmers must purchase, first by creating hybrid seeds that are less productive after the first generation, then creating legal protections called plant breeders' rights (PBR), and finally getting the right to patent GE seeds (Kloppenburg [1988] 2005).

Biotechnology patents, first granted in the 1980s, recognize transgenic materials and organisms as "inventions." One of the most significant outcomes of the development of intellectual property protections for genes and organisms has been the concentration of the seed industry. Encouraged by technological breakthroughs and the emerging intellectual property rules, multinational chemical companies, global pharmaceutical companies, and small start-up biotechnology firms raced to develop commercial applications of genetic engineering. In order to acquire patents held by smaller companies, the major agrichemical firms simply bought the smaller companies. Then, discovering that they needed access to seed distribution systems, they proceeded to acquire prominent seed companies, creating the high levels of concentration that we see today (Boyd 2003).[10]

While GE crops emerged out of a particular set of historical circumstances, questions about food safety, environmental integrity, and economic impacts have prompted a plethora of new institutional changes in the organization of the farm input industry, food distribution systems, and regulatory systems at local, national, and international levels (Schurman 2003). Many of these changes—such as the formation of consumer labels, traceability systems, and international treaties on trade in GE organisms—are explored in the chapters to come. Driving many of these changes are concerns about health and environmental safety. Although there is not adequate space here to fully discuss these concerns, a short overview provides a glimpse into the debates.

Critics of GE crops, and those who advocate caution in commercializing them, have identified possible food safety concerns, such as the potential for introducing new allergens into foods, the medical consequences of using antibiotic-resistance genes in the genetic engineering process, and the prospect of inadvertently increasing the level of toxins in plant materials (Union of Concerned Scientists 2002; Center for Food Safety 2000). Some published research indicates that GE foods may have negative health impacts (Dona and Arvanitoyannis 2009; Ewen and Pusztai 1999). These studies have been enormously controversial, with supporters and detractors battling over the credibility of each claim. Scientists have also pointed to concerns about the inputs associated with GE crops, such as the Roundup Ready trait that allows farmers to spray glyphosate throughout the growing season to control weeds. In June 2011, a group of concerned scientists published a review of studies suggesting that glyphosate may be linked to birth defects (Antoniou et al. 2011).

Advocates of biotechnology maintain that there is overwhelming scientific evidence that GE foods are safe to eat and regulatory agencies are doing a good job protecting the public. In a 2004 report, the National Research Council (2004, 8) indicated that "no adverse health effects attributed to genetic engineering have been documented in the human population." At the same time, the National Research Council highlighted the inadequacy of current knowledge and technology to predict the health effects associated with compositional changes in GE foods (ibid., 10). It is generally the responsibility of regulatory agencies, such as the Food and

Drug Administration (FDA) in the United States, to ensure that GE foods that pose known safety threats are not commercialized.[11] Regulatory procedures, however, may not be adequate to prevent unanticipated health consequences, due to limitations of the scientific evidence available about the differences between transgenic and conventional crops (Pelletier 2005, 2006; Billings and Shorett 2007; Millstone, Brunner, and Mayer 1999).

Another key area of research about the impacts of GE crops deals with environmental consequences. A position paper published by the Ecological Society of America (a professional scientific organization) in 2005 indicated that the possible negative ecological effects of releasing GE organisms into the environment could include, under certain circumstances:

(1) Creating new or more vigorous pests and pathogens; (2) exacerbating the effects of existing pests through hybridization with related transgenic organisms; (3) harm to nontarget species, such as soil organisms, non-pest insects, birds, and other animals; (4) disruption of biotic communities, including agro-ecosystems; and (5) irreparable loss or changes in species diversity or genetic diversity within species. (Snow et al. 2005)

Each of these areas of concern is a subject of ongoing research among ecologists, population biologists, crop scientists, and other investigators. As two astute participants in this research field point out, "Scientific understanding of the factors affecting environmental risk is still developing, and it is unlikely that controversies over environmental risks of genetically engineered organisms will be resolved in the near future" (Andow and Zwahlen 2006).[12]

As indicated earlier, this book focuses on social conflicts over the release of GE crops into environments where they are free to reproduce, resulting in "genes out of place." As anthropologist Birgit Müller (2006) puts it, "Although engineered by man to serve human purposes, from the moment onward when genetically engineered plants are released into the environment they escape human control and develop their own agency"—that is, they behave like any other living plant, growing, reproducing, and spreading their genetic material to offspring. Cases in which environmental releases of GE crops led to unwanted food or farm contamination are becoming increasingly familiar. In Hawaii, for instance, a GE variety of papaya was introduced in 1998 in order to combat a virus that was devastating the crop. A few years later, a coalition of activists discovered that GE papaya seeds had become widely dispersed throughout the seed supply, thwarting some farmers' desires to produce non-GE fruit (Bondera and Query 2006). In Australia, canola repeatedly tested positive for GE material even when GE seeds were prohibited, perhaps because of experimental field trials or mislabeled seed bags (Australian Broadcasting Corporation 2005, 2006). And 2009 brought a series of puzzling discoveries of food containing a type of GE flax that was officially deregistered and destroyed in 2001, yet evidently remained in the Canadian seed supply (Greenpeace International and GeneWatch UK 2009, Nickel 2010). Ecologists use the term *transgene flow* to refer to the process through which genes from GE crops are incorporated into the gene pool of another plant population, such as wild relatives or conventional (non-GE) crops. The consequences of transgene flow are varied, depending on an array of social and biological factors. Transgene flow can occur through the dispersal of seeds or propagules (e.g., buds, roots, or leaf cuttings) or the diffusion of pollen. For wild plants, the impacts depend on whether the resulting hybrids (genetic mixtures of GE and non-GE plants) are better suited to survive than the rest of the wild population. In some cases, wild hybrids may have a lower fitness, so the wild plant population will shrink. In other cases, they may have a higher fitness, becoming invasive—a so-called superweed. For agricultural crops, the impacts documented in some circumstances are reduced seed quality, threats to food safety, violations of organic standards for agriculture, and offenses to cultural values (Andow and Zwahlen 2006).[13]

Science and Social Movements

As the above discussion suggests, concerns about transgene flow are not limited to health or environmental issues; there are also farmers and consumers who wish to avoid genetic "contamination" because they adhere to farming principles that exclude biotechnology or find GE crops culturally inappropriate. In this book, I refer to these kinds of farming efforts as "alternative pathways" in agriculture, such as organic farming, localism, autonomy from transnational corporations, and the renewal of traditional farming practices and indigenous knowledge. I borrow this phrase from David Hess (2007, 4), who uses it to describe collective action to build alternative institutions and change cultural practices. Organic farming, sustainable rural development projects, efforts to renew indigenous agricultural traditions, alternative seed exchanges, and local food movements could all be considered alternative pathways in agriculture. In some efforts to pursue and maintain these forms of alternative agriculture, farmers actively seek to limit the reach of the biotechnology industry, forming alliances with environmentalists and the transnational antibiotech movement.

Antibiotech activists have been increasingly attentive to incidents of genetic contamination since the late 1990s. In 2005, two organizations, Greenpeace International and GeneWatch UK (2011) launched an online database "to record all incidents of contamination arising from the intentional or accidental release" of GE crops. The concern voiced by these two organizations is that "once released, it [is not] possible to contain or control these organisms yet there is no global monitoring system." Beyond documenting contamination, these and other organizations challenge the public policies, research agendas, and agricultural practices that have facilitated the rapid expansion of GE crop cultivation.

Probiotech commentators often demonize those who challenge the benefits of transgenic crops as "antiscience zealots" or irrational "deniers" of science. For instance, one recent book simplistically characterizes the debate about GE crops as a "battleground on which the forces of science and anti-science now clash" (Taverne 2005, 7). In the foreword to another book on the subject, Nobel Prize winner Norman Borlaug and former US president Jimmy Carter suggest that overregulation of GE crops "in Europe and elsewhere might have been avoided had more people received a better education in biological science" (Paarlberg 2008, ix).[14] In the chapters to come, I hope to challenge and complicate this portrayal of activists as antiscience by probing the ways that antibiotech activists interact with, participate in, and contest the "science-based" institutions that govern transgenic crops.[15]

Sociologists have offered many different definitions of a social movement. Some definitions are state centered, indicating that the term social movement only applies to collective action that confronts the government. In this book, however, I apply a broader definition. A social movement is a conscious, collective attempt to confront powerful opponents in order to create cultural, political, or economic change. Social movements not only seek policy change or political inclusion but also may seek symbolic change in institutions or culture, or changes to the "rules of the game" that form the basis of political action and everyday life. Recently, this understanding of social movements has been dubbed a "multi-institutional politics approach" (Armstrong and Bernstein 2008). Research in this vein is attuned to challenges to culture (including cultural ideas about science) as well as the state and other institutions. Sociologists Elizabeth Armstrong and Mary Bernstein (ibid.) argue that analysts of social movements should not assume, a priori, that national governments are central targets for activism, although they frequently are. The strategies and targets that social movements choose, and the outcomes of their confrontations with different institutions, reveal the complex ways that power is structured in a society. By studying social movements, therefore, we learn about the overlapping and often contradictory institutions that structure the social world.[16]

In part, opposition to GE crops is a struggle over material resources. Activists seek to defend the livelihoods of small farmers and protest the concentration of wealth in the hands of biotech companies. These struggles are, at the same time, battles over meaning, classification, and cultural rules. Can transgenic plants be considered a form of pollution? Should genes be considered the property of biotechnology companies? Is science a suitable basis for deciding whether to introduce GE crops to the market? Who is an expert in debates about biotechnology? In other words, antibiotech activists are not simply protesters against a technology or industry. Throughout its history, the antibiotech movement has generated and disseminated new knowledge claims, transforming popular understandings of the implications of GE crops for society and the environment while also reshaping the categories used to govern them.

Given the uncertain nature of the threats posed by releasing GE organisms into the environment, and the prominence of environmental and safety critiques of biotechnology, the "risk society" thesis developed by the well-known German sociologist Ulrich Beck is a prevalent approach to understanding public apprehension about GE crops. Beck (1992, 1995) suggests that as an outcome of modernization, contemporary industrial societies have experienced the proliferation of new risks. Hazards resulting from new industrial processes and technological developments are difficult to measure and quantify, despite the prevailing adherence to presumably calculable risk-benefit decision making. Indeed, risks like radiation, cancer-causing toxins, and greenhouse gases are often invisible to observers; the public is therefore increasingly dependent on scientists to characterize the nature of the risks they face. Yet in the risk society, the place of science and technology—widely seen as the source of these novel risks—is precarious. Science typically produces incomplete and contradictory knowledge about contemporary hazards. As a result, there is growing public criticism of the institutions of science and a distrust of scientific experts, especially among environmentalists.

The risk society approach suggests that people may fear the new GE foods because they do not trust the expert institutions that produced and govern them. In many respects, conflicts over GE crops and genes out of place do appear to bear out Beck's insights about risk society. Many of the early critics of GE crops discussed the technology in terms of the anticipation of novel dangers. Over the years, the risk-oriented approach to GE activism resonated with international environmental politics and public concerns about food safety. Opponents of GE crops have, in this context, frequently drawn parallels with other risky technologies when considering the release of GE crops into the environment, as evidenced in this quotation from Canadian geneticist and environmental advocate David Suzuki (2000):

We only have to reflect on DDT, nuclear power, and CFCs, which were hailed as wonderful creations but whose long-term detrimental effects were only found decades after their widespread use. Now, with a more wise and balanced perspective, we are cutting back on the use of these technologies. But with genetically modified (GM) foods, this option may not be available. The difference with GM food is that once the genie is out of the bottle, it will be difficult or impossible to stuff it back. If we stop using DDT and CFCs, nature may be able to undo most of the damage—even nuclear waste decays over time. But GM plants are living organisms. Once these new life forms have become established in our surroundings, they can replicate, change, and spread; there may be no turning back.

Critics like Suzuki challenge the claims that GE crops are safe, either for human consumption or natural ecosystems, noting flaws, uncertainties, and gaps in the scientific research. They also point out the open-ended nature of the risk: "There may be no turning back."[17] The risk society thesis thus seems to explain objections to GE crops.

While insightful in many respects, the risk society framework understates the importance of social movements in translating scientific knowledge into political ideas and making risks meaningful to the public. Analyses of GE crop opposition that build on the risk society thesis tend to emphasize the characteristics of GE crops themselves, especially the scientific uncertainty surrounding their environmental and health effects, as the motivation for public protest. Public opposition to GE crops does not result directly from the introduction of a potentially risky technology, though. Rather, the risks—as studied and measured by scientists—are framed and made meaningful through the efforts of social movements. As sociologist Andrew Jamison (1996, 224) puts it, "It is not sufficient that facts and concepts are created by scientists in relation to 'natural' phenomena; global environmental problems also require intermediary actors with the ability to translate the calculations, simulations, and projections [of scientists] into issues of public concern." Jamison and his colleague Ron Eyerman, a sociologist, indicate that social movements are shapers of consciousness and public spaces in which knowledge is produced. That is, social movements do collective "knowledge work," producing new understandings of the world around them (Eyerman and Jamison 1991).

Rachel Schurman and William A. Munro (2006) demonstrate the utility of Eyerman and Jamison's insights for understanding the early antibiotech movement. Schurman and Munro trace the origins of antibiotech activism to two groups of people who initially raised distinct sets of concerns. In the 1970s and 1980s, "critically minded scientists, environmentalists, and technology skeptics" in the United States and Europe asked questions about "the dangers of this novel technology to human beings and other living things and the social, moral, and ethical issues raised by intervening in nature with such a powerful new set of tools" (Schurman and Munro 2010, 57). At roughly the same time, "development critics"—researchers and activists whose work questioned the premises and purported benefits of economic modernization schemes for the Global South—formed a critique of biotechnology that originated with concerns about corporate control over seeds and the erosion of plant genetic diversity. Throughout the 1980s and early 1990s, "as these individuals' work, social concerns, and worldviews intersected, they began to form a coherent analysis of the technology" (ibid., 2010, 65). Present-day critiques of GE crops emerged, in large part, from the early efforts of these critical scientists and activists to make sense of the new tools of biotechnology.

As a result of this diverse intellectual history, for many critics of biotechnology, anxieties about the environment and health are entwined with concerns about power and inequality in the global agrifood system. From the critical perspective offered by observers like Vandana Shiva (a prominent Indian activist) and Jack Kloppenburg (a US sociologist), the social and biological dimensions of GE crops are not separate problems but rather parts of the same process of increasing corporate control over agriculture and natural resources. In a 1995 essay on biotechnology and the conservation of biodiversity, Shiva (1995, 198–199) writes:

Ecological erosion and destruction of livelihood are linked to one another. Displacement of diversity and of people's sources of sustenance both arise from a view of development and growth based on uniformity created through centralized control. In this process of control, reductionist science and technology act as handmaidens for economically powerful interests.

Making a similar argument, Kloppenburg ([1988] 2005, 314; emphasis added) explains his views about the possible beneficial uses of biotechnology:

I believe that biotechnology in general—and genetic engineering in particular—might be used in safe, socially progressive, and sustainable ways, if social circumstances allow them to be developed in a manner appropriate to those goals. . . . *Ecological and social sustainability will follow principally from the social arrangements we construct, not from the technologies we create.* . . . So, while I do not believe that we now possess the scientific knowledge to use the new genetic technologies safely, I am even more concerned that we do not now have social institutions in place to see that they are used properly and well. . . . The powerful tools of biotechnology are now being wielded largely by a narrow set of corporations which claim to want to use them to eliminate hunger, protect the environment, and cure disease, but which in fact simply want to use them as quickly as they can to make money just as fast as possible.

Shiva and Kloppenburg draw attention to problems that are often obscured by the dominant "risk" debate about GE crops, offering not just a critique of the possibility of harm but also the social systems that allow those harms to proliferate.

Social critiques like these are rarely incorporated into formal regulatory processes. Indeed, as I indicated in my discussion of the GE alfalfa conflict, advocates of GE crops have vehemently opposed allowing the government to consider anything but purported science-based assessments in regulatory decisions. And as I will explain in chapter 2, the governance of GE crops has also been scientized in global agreements such as the World Trade Organization. The cases in this book, however, do not present a simple dichotomy of "local social values" versus "global technocratic institutions." Social movement interpretations of the problems of GE crops are seldom in complete contradiction to the dominant discourses about them. As anthropologists James Fairhead and Melissa Leach (2003, 2) explain, in environmental governance, global and local forces

can be united by shared problem-framings: the local concerns or forms of knowledge which come to be represented in national and international debates frequently share dominant, globalised questions (for instance, about what is happening, and which trends need to be arrested or modified), while alternative framings and their implications drop out of view. . . . In this sense, even the most distantly "local" actors and their everyday lives may be caught up in the vortex of global debate—although with highly unequal capacities to shape its terms.

One of the questions that motivate this study is how and under what circumstances those who defend alternative forms of social life—namely, alternative pathways in agriculture—are able to shape the terms of the "vortex of global debate."

Advocates of alternative agriculture frequently draw upon the dominant discourse of scientific risk assessment to express their opposition to GE crops. Given the widespread cultural authority of science, it is not surprising that social movements incorporate science into their tactics, such as using counterexperts, publicizing suppressed studies, and carrying out their own participatory research. In many social movements, scientists are crucial partners with those seeking social change. Scientific actors and institutions contribute to defining emerging issues, provide information to activists, and take part in political conflicts at both the domestic and international level. The relationship between science and social movements is perhaps most pronounced (and most widely analyzed) in studies of environmental movements, which have long relied on science to press the importance of their concerns (Hays 1987; Schnaiberg 1980; Dunlap and Mertig 1992). Other examples, though, indicate that science and activism merge across a wide variety of historic and contemporary movements. Social scientists served as expert witnesses in landmark civil rights lawsuits in the United States (Jackson 2001), and today, antiracism campaigns use statistics on African American incarceration, analyzed by sociologists (Oliver 2008). In another example, gay rights movements have at times grappled with the scientific possibility of a "gay gene," and considered controversial research on brain differences between gay and straight people (Brookey 2002).

Sociologists of science and others in the interdisciplinary field of science and technology studies have been observing these trends for some time. Much of the science and technology studies work on social movements has focused on science and technology as the arenas of contention, asking how activism changes scientific knowledge and practice (for a review, see Hess et al. 2008). Social movements can bring about new ideas and methods in scientific research (Epstein 1996) or technological development (Hess 2007). Mobilized citizens can become scientific experts and knowledge producers themselves on matters of significance to them, such as illness or local environmental contamination (Brown 1992, 2007; McCormick, Brown, and Zavestoski 2003; Hess 2009; Corburn 2005). Activists, particularly in the health arena, can have a significant impact on the causal claims that scientists make. For example, a grassroots self-help group helped transform how medical professionals understand the causes of endometriosis (Čapek 2000).

Similarly, the antibiotech activists in this book seek to change the way that knowledge about biotechnology is produced. Yet they also fight to ensure that expert discourse does not overshadow citizens' perspectives on environmental, social, economic, and moral issues in decisions about scientific and technological developments. Comparable efforts to offer alternative framings of ostensibly technical questions have been observed in a variety of cases. For instance, grassroots activists and their transnational allies have spotlighted the human rights abuses associated with big dam projects—in the process challenging the very premise of "development"

(Khagram 2004). Elsewhere, indigenous peoples have fought for access to clean water, demanding that regulators recognize that water is not only a health and environmental matter, as commonly understood, but also sacred to indigenous people (Espeland 1998; Mascarenhas 2007). And across the United States, community struggles against environmental racism have pointed out the racial injustice at the root of environmental pollution, bringing issues of inequality to the forefront of conflicts over the environment (Bullard 2005). The impact of these kinds of struggles, when successful, is to reverse the scientization of social problems, reopening the possibility of democratic deliberation about the value preferences and power relations that were previously obscured. When citizens mobilize to challenge the narrowly technical definition of a policy issue, they politicize (and potentially democratize) decision-making processes. The belief in science as an objective, nonpartisan source of knowledge may remain unscathed, but the relative importance of science in making policy and regulatory decisions is diminished, as social values—and the people who voice them—gain a more central role in the political process.

In the case studies presented in the rest of this book, I examine four main strategies that farmers and activists have used in reaction to the scientization of biotechnology politics: externalizing their struggles to international experts, carrying out civil society research, scrutinizing science in court, and using market-based tactics. These strategies certainly do not exhaust the list of ways that social movements may challenge technology and question the political authority of science. However, they point to diverse ways that advocates of alternative pathways in agriculture are reshaping the rules of the global debate about GE crops.

Cases and Research Methods

In this book, I analyze how antibiotech activists interact with, participate in, and challenge the institutions that govern transgenic crops. Looking at controversies surrounding two different crops in two different countries, I examine a variety of social movement strategies to protect alternative agricultural practices from GE contamination. These episodes of contention reveal not only the extent to which biotechnology politics is scientized in different countries and institutional contexts but also the possibilities for social movements to advance alternatives to risk-oriented assessments of new technology. The first case is in Mexico, where a controversy over planting GE maize (corn) has been raging since the late 1990s. The second case is in Canada, where the contamination of canola crops with GE varieties sparked two precedent-setting legal struggles that pitted farmers and environmental activists against the biotechnology industry.

In Mexico, rural communities and environmental activists protested vehemently when native maize varieties were found to be contaminated with GE varieties imported from the United States. A transnational network of activists mobilized to protest the importation of GE corn and demand the protection of native Mexican maize. The network of protesters involved a diverse array of groups, including regional rural organizations, environmental organizations based in Mexico City, international nongovernmental organizations (NGOs), and an assortment of committed scientists and intellectuals around the world. In describing what is at stake in the conflict over GE maize in Mexico, geographer Kathleen McAfee (2003, 18), a perceptive observer of the conflict, summarizes the issues this way:

The possible risks posed by traveling transgenes are not well understood, but there are plausible scientific reasons for concerns about possible hazards to agricultural biodiversity and agro-ecosystems. More troubling, however, are the likely consequences—for local food security, cultural survival, and national economic sovereignty—of the private ownership of staple-crop genetic resources and of the influence on trade policy, agricultural research, seed and food markets, and farming-system options of a small number of powerful states and transnational firms.[18]

Organic and conventional farmers in Canada faced legal and economic consequences when GE canola infiltrated their fields. In one prominent case, the biotechnology company Monsanto sued a Saskatchewan canola farmer for cultivating patented herbicide-resistant canola that arrived in his fields as the result of cross-pollination. In a separate case, a coalition of organic farmers pursued a class action lawsuit on behalf of Saskatchewan's fifteen hundred certified organic canola producers against Monsanto and Aventis, claiming that that the two corporations knowingly released genetic pollution when they marketed GE canola. Legal scholar Martin Phillipson argues that these two cases are closely related. The decision in the former one, notes Phillipson (2001) "provided ample evidence of the array of legal rights enjoyed by agricultural biotechnology corporations and associated TUAs [Technology Use Agreements]. However, the decision did not address the essential corollary of legal rights: legal obligations," which were the subject of the second case. In both instances, farmers and their allies in the anti-biotech movement battled the biotechnology industry in court over the legal status—and social consequences—of transgenic material in the environment.

This is not the first study to look comparatively at the politics of biotechnology. There has been a great deal of research investigating the diverging experiences of the European Union and United States with GE crops (Bernauer 2003; Jasanoff 2005; Prakash and Kollman 2003; Murphy and Levidow 2006). The European Union and United States are the most powerful players in the international politics of GE crops, thereby warranting the extended attention they have received. This study, however, turns the spotlight on farmers and rural communities in countries whose approaches to GE crop governance are less widely scrutinized or understood. This analysis differs from other comparative studies of biotechnology policymaking in two other ways as well. First, I focus on the attribution of meaning to genes out of place. That is, rather than looking broadly at the formal policies and processes for governing GE crops, I concentrate on particular episodes of conflict over how transgenes in the environment are to be governed. My comparison also does not necessarily center on the national state or distinctions between national cultures. As the chapters to come will explore in detail, struggles to define genes out of place in Mexico unfold in different ways than parallel conflicts in Canada. This is not only because of cultural differences between the two countries or farming communities but also, significantly, because of their specific locations within global relationships of trade and commitments to international environmental treaties.

I use the concept of genes out of place to indicate that the importance of transgenes in the environment is a matter of contestation. The genes out of place notion draws on the insights of Mary Douglas, a twentieth-century anthropologist whose studies of purity and pollution have been highly influential in the social sciences. In a classic anthropological text, Douglas observes that dirt is simply "matter out of place." Soil in the garden is dirt on the kitchen floor. A tasty dinner becomes filth when spattered on a shirt. Douglas ([1966] 1978, 2) suggests that ideas about dirt and pollution reflect "cherished classifications," and often reveal crucial characteristics of the social order.

As we know it, dirt is essentially disorder. There is no such thing as absolute dirt: it is in the eye of the beholder. If we shun dirt, it is not because of craven fear, still less dread of holy terror. . . . Dirt offends against order. Eliminating it is not a negative movement, but a positive movement to organise the environment.

Douglas's insights help to make sense of cultural boundaries, from religious dietary rules (a subject of Douglas's own research) to the idea of pure science. If notions about contamination and purity are socially constructed, and highly variable across cultures, then it should be possible to analyze how those categories come into being, are institutionalized, become points of conflict, and change.[19]

With respect to transgene flow, there is no such thing as "absolute contamination"—a distinction between purity and pollution that is self-evident to all. To be clear, I am not questioning the existence of transgenes or transgene flow, or the possibility of scientifically documenting their effects. What concerns me here is how different groups of people view genes out of place, and what conflicts over those diverging perspectives tell us about the social order. Critics of biotechnology typically contend that genes out of place constitute contamination or "biopollution." In one of the cases examined in this book, Mexican opponents of GE corn characterized transgene flow as contamination and demanded that the state take action to prevent pollution of the native maize crop. Advocates of GE crops framed the phenomenon differently, typically using more benign-sounding words like "admixture" or "adventitious presence." This is seen in another case explored in this book, in which biotechnology companies argued that they should have no legal liability for the adventitious presence of GE canola appearing in Canadian organic crops. These differences in understanding are not only scientific disagreements about how to act on the uncertainty surrounding the impacts on the environment or health, nor are they simply matters of personal interpretation or individual preference. Conflicts over genes out of place are *political struggles over the social order*—more specifically, over the social organization of agriculture. In a global context that is marked by both ecological diversity and economic inequality, transgene flow has different ramifications for people depending on their social position.

In each episode of conflict I examine here, the outcome of the struggle to define genes out of place had material consequences. In one instance, an international body of experts, tasked by the Commission for Environmental Cooperation to examine the issue of GE maize in Mexico, concluded that genes out of place were so laden with symbolic meaning that decision making must be made more democratic. In another example, Mexican farmers and environmentalists understood genes out of place as potentially dangerous environmental pollution and undertook an effort to monitor them closely. When a Canadian court ruling found that genes out of place were inventions, covered by a patent held by the company Monsanto, a farmer was held liable for patent infringement. In another Canadian legal decision, genes out of place were assumed to be benign changes to the food supply and organic farmers were denied an opportunity to pursue a class action lawsuit to remedy the losses associated with contamination of organic crops. In each case, opponents of GE crops aimed to redefine genes out of place as contamination—an unwanted and harmful intrusion—and to institutionalize that idea in laws, international environmental agreements, and the "common sense" of scientists and regulators. Whether and how they succeeded tells us a great deal about the particular configurations of power in each national context—which reach well beyond government into the realms of science, industry, and the international policy arena.

Indeed, unlike Douglas's anthropological case studies, the cultural contexts in which genes out of place become socially meaningful are not closed societies with clear boundaries. My aim is not to study the cases in isolation in order to draw out independent variables and produce generalizations. I am attentive to the fact that the examples I study—while occurring in two different countries—are not independent of one another. Antibiotech activists around the world are linked through transnational advocacy networks and share organizational resources. Furthermore, scientific knowledge about transgene flow is created and shared globally, becoming a vital resource to those who protest against GE crops as well as those who advocate for biotechnology. Finally, national policies themselves are influenced by international trade relations, treaties, and laws. Far from being isolated instances, bounded by national borders, the politics of genes out of place in each situation are markedly influenced by the policies and regulations of dominant global actors.[20] The character of the debate in each case is significantly affected by the location of each group of agricultural producers in relation to global food commodity chains. Understanding these global systems is essential to making sense of the extremely heated and contentious responses to genetic engineering that have emerged around the world.

I collected data on the struggles over GE maize and canola through interviews, participant observations, archival research, and analyses of published documents (e.g., legal documents, scientific papers, newspaper articles, and activist Web sites). I interviewed approximately eighty individuals, including representatives of civil society organizations, research scientists (both critical of and favorably disposed toward genetic engineering); independent scholar-activists, representatives of Mexican and Canadian regulatory agencies, Mexican maize producers, Canadian organic farmers, and representatives of the biotechnology and canola industries.[21] I conducted interviews in English and Spanish. The interviews generally lasted between one and two hours, and were recorded and transcribed.

Although each of these interviews shaped my understanding of these cases, I do not reproduce quotations from all of them in this book. Instead, I use a combination of information gleaned from interviews, documents, and secondary sources to reconstruct the history of each case. Where I do quote the people I interviewed, I conceal their identities in order to protect their confidentiality. In a small number of instances, I use the real names of my informants, with their permission, when they are pivotal figures in particular conflicts.

My field research in Mexico took place from September 2005 to April 2006. In Mexico City and Oaxaca, I interviewed activists, farmers, intellectuals, researchers, plant breeders, and government officials involved in the debate over transgene flow in maize. I also learned about the GE maize controversy through participant observation, working as a volunteer at a small environmental organization in Mexico City, and attending a wide array of meetings and workshops in Mexico City and Oaxaca. These ranged from gatherings organized by the Mexican government's environmental agency, Secretaría de Medio Ambiente y Recursos Naturales (SEMARNAT), on the topic of biotechnology and biodiversity protection to a national forum of rural organizations participating in the "network in defense of maize."

The bulk of data collection on the conflicts over Canadian canola took place in the province of Saskatchewan in July 2006. I interviewed key actors in the two lawsuits that were at the center of contention over transgene flow in Saskatchewan. I also interviewed several representatives of the biotechnology industry, canola industry, and Canadian Wheat Board, and a small number of public officials and scientists with varied perspectives on the topic. I briefly became a participant observer in the network of organic farmers, attending two organic "field days"—events in which organic farmers visit organic farms in their area to learn from one another about the techniques they use and problems they face. Finally, I gathered and analyzed legal documents connected to the two lawsuits that are at the heart of the struggles over GE canola. After my fieldwork was completed, I continued to follow the controversies over GE maize and canola through activist email lists, media coverage, and periodic contact with key informants.

Chapter Overview
Conflicts over genes out of place are, at their core, political struggles over the shape of the global agrifood system. A feature of these struggles is a debate about whether decisions are being made scientifically or not.

Chapter 2 develops this line of argument more fully, while also introducing the political context in which the controversies over GE maize in Mexico and GE canola in Canada emerged. I argue that the combined processes of *neoliberalization* and *scientization* constrain opportunities for public participation in decision making about agriculture. Governments around the world have embraced "free market" principles for the governance of agriculture. At the same time, the political and economic implications of biotechnology have been sidelined in favor of a science-based assessment of health and safety risks. In both Mexico and Canada, the combination of these shifts has left the state with a responsibility to govern certain natural resources while minimizing debate or discussion about the organization of society and the market. I examine the implications of these combined changes in governance, both globally and in each of the two countries, laying out the context for the conflicts examined in the subsequent chapters.

Chapters 3 and 4 focus on what I broadly refer to as Mexican "maize activism." When US scientists discovered transgenic material in samples of maize from rural Oaxaca, there was an immediate firestorm, both about the credibility of the study and the ecological consequences of introducing GE crops to a place considered to be a vital source of maize genetic diversity. In chapter 3, I discuss the emergence of a movement "in defense of maize," and its relationship with an international environmental organization, the Commission for Environmental Cooperation. I introduce the concept of an "epistemic boomerang," a strategy of mobilization in which local groups and NGOs, frustrated by exclusion from policymaking debates, go outside their normal political channels to appeal to scientists, thereby hoping to mobilize scientific research in support of their social goals.

Chapter 4 concentrates on an attempt at activist-led environmental monitoring, aimed at determining the extent of transgenic contamination of native Mexican maize. In this chapter, I explore environmental monitoring as a social movement strategy for marginalized groups in the Global South. I build on previous studies that examine the processes through which science and technology are opened up to public scrutiny, and consider the effects of activist-led monitoring efforts on the formation of an activist network that includes peasant farmers, indigenous peoples, NGOs, concerned scientist organizations, and transnational advocacy groups.

In chapters 5 and 6, I look at Canadian controversies over GE canola. Chapter 5 focuses on the question of patents on transgenic material released into the environment. I examine the unique characteristics of the legal system as an institution in which to change the global debate about GE crops. When Monsanto sued a farmer, Percy Schmeiser, for allegedly using, without permission, the company's patented herbicide-resistance gene, it prompted a public debate about the commodification of plant genetic resources—an issue that a small network of activists has been working on since the late 1970s, and now is one of the central concerns associated with genes out of place. This case, however, reveals that even debates about property rights and seed saving—eminently political topics—have become scientized, making it difficult for farmers, activists, and other nonscientists to have a substantial role in shaping public policy.

Chapter 6 considers a case in which organic canola producers, facing contamination of their crops, pursued a class action lawsuit against two biotechnology companies, accusing them of contaminating their crops to such an extent that organic canola became unmarketable. This case demonstrates an important pattern in the governance of GE crops, in which consumers abroad reject GE foods, leading farmers to press seed producers into removing transgenic seeds from the market. Voluntary, market-based governance of GE crops has serious limitations, however, and over the last decade, farmer groups and antibiotech activists have sought to strengthen the role of the state, calling for a criterion of "marketability" to included in the regulation of GE crops. However, as I show in this chapter, the Canadian judicial, regulatory, and legislative systems have been strongly resistant to this idea, echoing industry demands for "science based" regulation.

The case studies in this book reveal the difficulty of establishing the legitimacy of political considerations and alternative sources of knowledge in the governance of biotechnology. Nevertheless, I find that struggles over the social consequences of genes out of place expose diverse and creative challenges to the existing social order. In the conclusion, I discuss the four main strategies investigated in this book—externalizing the struggle to international experts, carrying out civil society research, scrutinizing science in court, and using market-oriented tactics—and consider their applicability to other social struggles. Then I return to the question raised in the book's opening pages: What are the consequences of reducing political conflicts to scientific and risk-oriented questions, and how might we do things differently? Like many of the activists featured in this book, I argue that the social and ecological consequences of GE crops are too important—and too diverse—to be left either to markets or experts alone.

2

Free Markets, Sound Science

Canada and Mexico are two very different countries, with distinct cultures, agricultural traditions, landscapes, and levels of economic development. Nevertheless, the opponents of GE crops in each country face remarkably similar resistance to the possibility of democratic debate about the social implications of biotechnology. Advocates of GE crops in both places frequently argue that decisions about whether to commercialize GE seeds and foods should be based on a scientific assessment of their effects on human health and the environment, not political considerations about how they will impact farmer livelihoods, rural communities, international trade relations, or the structure of the food industry. In this chapter, I contend that efforts to depoliticize the regulation of agricultural biotechnology reflect, on one hand, the growing dominance of neoliberal policy ideas in both the Global North and South, and on the other hand, an increasingly global process of scientization. By scientization, I mean the transformation of political conflict—in this case, over the desirability of transgenic crops—into a debate among scientific experts, ostensibly separate from the social context in which it unfolds. After discussing both of these concepts, I turn to the two countries that are the central cases in this book: Mexico and Canada. I describe the political terrain that constrains the debate about genes out of place, but that also offers some strategic opportunities for social movements to press for alternative forms of agriculture.

I use the term neoliberalism in this book to refer to a set of ideas and practices that have become, to a great extent, the taken-for-granted approach to the state-economy relationship in most countries around the world today. Geographer David Harvey (2005, 2) defines neoliberalism as "a theory of political economic practices that proposes that human well-being can best be advanced by liberating individual entrepreneurial freedoms and skills within an institutional framework characterized by strong private property rights, free markets, and free trade." In neoliberal theory, governments serve a variety of roles to enable the aforementioned framework. For example, as Harvey (ibid., 2) explains,

The state has to guarantee . . . the quality and integrity of money. It must also set up those military, defence, police, and legal structures and functions required to secure private property rights and to guarantee, by force if need be, the proper functioning of markets. Furthermore, if markets do not exist (in areas such as land, water, education, health care, social security, or environmental pollution) then they must be created, by state action if necessary. But beyond these tasks the state should not venture.

Neoliberal ideas are visible in efforts to privatize state functions, liberalize trade, and roll back social welfare programs. The roots of neoliberalism as a political ideology can be traced to a small intellectual movement in the 1970s, "stitched together from diverse strands in free-market economics, individualistic philosophy and anti-Keynesian politics" (Tickell and Peck 2003, 166) as well as "a deep, taken-for-granted belief in neoclassical economics" (Campbell and Pedersen 2001, 5). Neoliberal ideas spread globally throughout the 1980s and 1990s, in the form of "structural adjustment" programs—market reforms to encourage liberalized markets and foreign investment—which came to replace "development" as the conventional wisdom for the governments of poor nations, beginning in the 1980s (McMichael 2007). As sociologist Sarah Babb (2005, 200) notes, structural adjustment was a policy discourse explicitly developed by World Bank president Robert McNamara and cultivated by Western elites, "in keeping with the ascendant Reagan revolution." Governments were easily persuaded to adopt policy reforms in line with neoliberal ideas because of the outbreak of the third world debt crisis in 1982 and economic pressures to attract foreign investors. Many governments, including that of Mexico, also took advice from US-trained economists who strongly believed in neoliberal market reforms. The global prominence of neoliberalism as a discourse that shapes policy debates has grown since the 1990s with the establishment of free trade agreements and zones along with the World Trade Organization (WTO).

Neoliberal reforms in agriculture, including international free trade agreements, the deregulation of markets, the reduction or elimination of public subsidies and price support programs, and the privatization of agricultural research, have transformed global agrifood systems.[1] Trade agreements, such as the North American Free Trade Agreement (NAFTA), are based on the notion of comparative advantage—the idea that countries maximize their advantage in international trade by specializing in types of production that reflect the country's particular resources (such as cheap labor, plentiful sunshine, or access to laborsaving technologies). As I will discuss further below, for Mexico, pursuing comparative advantage has meant turning away from earlier policies that promoted domestic maize production to feed local urban and rural populations, in favor of a strategy of producing specialty crops for export (such as tomatoes) and dependency on the importation of inexpensive grain from the United States. As rural sociologist Philip McMichael (2007, 201) indicates, "Neoliberal food security means privileging food importing over local farming for many southern states." In combination with other neoliberal reforms, such as dismantling subsidy and price support programs, these changes have made food prices and farmers' incomes more vulnerable to market forces that are beyond their control.

Moreover, observes Harvey (2005), a key neoliberal idea is that where markets do not exist, they must be created. That is what the development of intellectual property rights for genetic material accomplished. Many governments have withdrawn from their historic role of developing and providing improved seeds for agriculture. International treaties, such as the WTO's Trade Related Intellectual Property Rights Agreement, have instituted the right to patent genetic sequences and transgenic organisms—thus creating new markets for these materials. As a consequence, the work of seed development has largely shifted to multinational corporations (Hubbard and Farmer to Farmer Campaign on Genetic Engineering 2009; Kuyek 2007b; Raustiala and Victor 2004).

With the development of markets for agricultural biotechnology, however, have come public concerns about its implications, driving the formation of new governance institutions and regulations, both globally and domestically. Neoliberalization thus is not simply a process of market creation and deregulation; it also involves new types of regulation "to address public concerns that could otherwise lead to a collapse of confidence in markets" (Moore et al. 2010). In the arena of biotechnology regulation, the question is not whether the technology will be regulated but rather how. What institutions will govern GE crops, and by what criteria? Many argue in favor of scientific criteria for decisions, saying that science clears away the tangle of interests and ideologies, enabling people to reach agreement. For instance, Nina Federoff, science adviser to the US state department, administrator of the Agency for International Development, and passionate advocate of GE crops, extols the virtues of science for talking "across chasms":

[Science] aspires to more democratic principles than many political systems because we have an external reference. People can have different theories, but we form an experiment to test it. It's the evidence that matters. So in science, we can have differences of opinion, but we can't have two sets of facts (Dreifus 2008).

Yet turning to science to create agreement means ignoring the wide-ranging reasons why one might be concerned about the trajectory of technological development. One scientist opposed to GE crops describes the situation this way:

The reasons for rejecting GE food are substantive and span multiple levels, from molecular genetics all the way to ecology and culture. Insistence on the right to—and need for—food free of genetic engineering is grounded in all these levels. It is unlikely that all of these arguments could ever be refuted by GE food proponents, which is presumably why most of these levels are so often excluded from official, industry-influenced debate. (Herbert 2005, 62)

The attempt to eliminate political considerations from the governance of GE crops is consistent with neoliberal ideas. Observers of neoliberal transformations point out that "one of the more far-reaching effects of . . . neoliberalisation has been the attempt to sequester key economic policy issues beyond the reach of explicit politicization" (Tickell and Peck 2003, 175). Agriculture, genetic resources, and food are certainly key economic policy issues, yet agricultural biotechnology is frequently treated as nonpolitical. Recall the argument, discussed in the previous chapter, by US elected officials against coexistence policies to protect non-GE alfalfa. They explicitly stated that regulatory decisions should be based on science, with other factors considered in the marketplace. In other words, the government should not intervene to favor particular styles of agriculture or types of farmers; instead, such matters should be worked out through science and free trade.

The linkage of neoliberal ideas with the notion of scientific risk assessment is perhaps most obvious at the WTO, the international body that deals with the rules of trade between nations. The WTO Agreement on the Application of Sanitary and Phytosanitary Measures (SPS Agreement) is one of the most globally significant policies on the evaluation of GE crops. The SPS Agreement was formed during the Uruguay Round of the General Agreement on Tariffs and Trade (GATT) negotiations (1986–1994)—the same negotiations that resulted in the WTO's formation. It was the first time that agricultural issues were brought into negotiations on global trade. The SPS Agreement does not specifically deal with GE crops but rather with food issues more generally. It aims to allow countries to protect the health and life of their consumers, animals, and plants against pests, diseases, and other threats to health, while preventing the use of regulations in an unjustified, arbitrary, or discriminatory fashion.

That is, the agreement allows countries to block imports of a food product that poses a health risk, but not to selectively block imports from certain countries while allowing others. To meet this objective, the SPS Agreement requires that the measures countries take either be based on scientific risk assessment or comply with the standards of one of three existing international bodies (which also use scientific risk assessment) (Winickoff et al. 2005; Winickoff and Bushey 2010; Gruszczynski 2006).[2] A government that believes that another country is violating the SPS agreement may bring a case before the WTO. For example, Canada and the United States recently won a lengthy battle against the European Union for its obstruction of trade in GE foods without "solid scientific arguments" to support its antibiotech position (European Stance on GMOs 2006).

There have been some prominent efforts to challenge WTO rules on trade in GE crops. In the 1990s, the United Nations held negotiations for the Biosafety Protocol, an international agreement that would address the ecological impacts of international trade in transgenic organisms. The protocol aims to "contribute to ensuring an adequate level of protection in the field of the safe transfer, handling and use of living modified organisms resulting from modern biotechnology that may have adverse effects on the conservation and sustainable use of biological diversity, taking also into account risks to human health." In contrast to the WTO, the protocol promotes the precautionary principle, referring to the 1992 Rio Declaration on Environment and Development. In that declaration, the precautionary principle is explained as follows: "Where there are threats of serious or irreversible damage, lack of full scientific certainty shall not be used as a reason for postponing cost-effective measures to prevent environmental degradation." In the Biosafety Protocol, application of this principle means allowing countries to regulate trade in transgenic organisms, in order to protect the environment, without clear proof that they cause harm. Yet in focusing on risks to the environment, the precautionary principle still places the central responsibility for governance on scientific experts rather than opening up democratic debate in order to resolve political conflicts about the direction of technological change in agriculture (Kleinman and Kinchy 2007).[3]

As my colleagues and I have written about elsewhere (Kleinman and Kinchy 2007; Kleinman, Kinchy, and Autry 2009), negotiators from a number of developing countries attempted to broaden the scope of the Biosafety Protocol, proposing language that would support the regulation of GE crops and other transgenic organisms on the basis of their expected socioeconomic impacts (see also Andrée 2007). Of course, the assessment of socioeconomic impacts could be scientized—carried out by experts without explicit political debate or input from the public. The proposals from the African Group and other nations of the Global South, however, highlighted the value conflicts and unequal social consequences associated with GE crops—topics not easily adjudicated through technical assessment. For example, the African Group proposed formalizing a process of socioeconomic risk assessment that would review:

a. Anticipated changes in the existing social and economic patterns resulting from the introduction of genetically modified organisms or GM products;
b. Possible social and economic costs of a loss of genetic diversity, employment, and market opportunities resulting from the introduction of GMOs;
c. Possible effects seen as contrary to the social, cultural, ethical and religious values of communities resulting from the use or release of GMOs. (UNEP Biosafety Working Group 1997, 95)

These proposals were not included in the document that was finally approved. Nevertheless, because the precautionary principle opens regulatory science up to public scrutiny, the Biosafety Protocol has offered some leverage to communities that are struggling to resist agricultural biotechnology.[4]

The distinction between science-based and precautionary approaches to governing the risks of GE crops, seen in the SPS agreement and Biosafety Protocol, is mirrored in the differences in US and EU regulatory frameworks. These regulatory approaches are often viewed as extreme opposites because the United States has been far more permissive of GE crops than the European Union has been.[5] There is significant variation among countries in the way that science and expertise are incorporated into political processes and public debate (Jasanoff 2005). As sociologist of science Brian Wynne (2007, 342) observes, though, despite the trade conflict and evident differences between the United States and European Union regarding transgenic crops, both share a "fundamentally similar transatlantic institutional discourse of sound science and risk." Wynne argues that in both Europe and North America since the 1950s, the sociocultural role of science has shifted—from *informing* to *defining* policy issues. This is most evident in the "pervasive discourse of *risk* as a way of defining public issues" (ibid., 344). Wynne argues that maintaining risk as a defining discourse for policy debate about science and technology "obscures public questions about upstream human purposes, ends, [and] visions driving innovation-oriented science" (364). For instance, for regulatory authorities, "the GE crops issue is, 'What are the risks (as defined by institutional science)?' rather than, 'What kind of agriculture do we want? Under what conditions could GE fit in, and are those conditions feasible and acceptable?'" (349). Democratic debate about the latter questions—and more generally, questions about social needs and priorities—are preempted by scientistic risk discourse.[6]

Wynne (ibid., 365) indicates that the ascendance of risk discourse is a "cultural syndrome" for which responsibility cannot be directly attributed. At least in part, the scientization of politics is a continuation of the long historical process of rationalization, first theorized by Max Weber in the early twentieth century. Weber (1991, 139) observed that the history of modern Western societies has been a historical drive toward a world in which "one can, in principle, master all things by calculation." This involves increasing control over nature along with a primary emphasis on rational action in the areas of the economy, law, and all other areas of society and culture. Following Weber, scholars often describe scientization as a broad and increasingly global trend, associated with advancing modernity (Beck 1992; Drori and Meyer 2006; Drori et al. 2003).

The scientization of social problems and politics has expanded over the past several decades (Drori and Meyer 2006; Drori et al. 2003). In the 1960s and 1970s, European and US political theorists began to note a shift toward a technocratic model of governance in which politics is replaced by a scientifically rationalized administration (Price 1965; Habermas 1970; Benveniste 1973). As prominent science policy scholar Sheila Jasanoff (1990) suggests, technical experts have become the "fifth branch" of the US government, providing advice and policy guidance while maintaining the appearance of political neutrality. Much of the scholarship documenting the scientization of politics focuses on the United States and Europe, but there is growing evidence that the trend has also reached other parts of the world (Drori and Meyer 2006; Kinchy, Kleinman, and Autry 2008; Kleinman, Kinchy, and Autry 2009). To give one example, a recent study showed that Brazilian dam-building policy, shaped by "European or American-educated technocrats," is also scientized, as "decisions about regulating dam building are based on environmental impact assessments generated by hired scientific consultants" (McCormick 2006, 327).

When social policy and regulation are primarily shaped by science and technocratic decision making, this frequently serves to promote industry's interests. Critics of the sound science argument make the point that

the insatiable quest for "better science" in policymaking has become a significant and powerful tool used to support dominant political and socioeconomic systems. Through this "scientization" of decision making, industry exerts considerable control over debates regarding the costs, benefits, and potential risks of new technologies and industrial production by deploying scientific experts who work to ensure that battles over policymaking remain scientific, "objective," and effectively separated from the social milieu in which they unfold. (Morello-Frosch et al. 2006, 245)

Through scientization, political and moral questions become either inappropriately framed in scientific terms or simply sidelined in the mainstream debate. Calls for sound science and risk discourse delegitimize the importance of so-called nonscientific issues while limiting public participation in decision making. This gives industry, with its assurances of scientific risk assessment, a distinct advantage. In the case of GE crops and other instances of present-day political conflict, there are indications that scientization is a strategic political project, pursued by actors who stand to gain by constructing matters of social significance in a narrowly technical way. It will be clear in the chapters to come that proponents of GE crops strategically advocate science-based and risk-oriented regulation in order to prevent an increase in government regulation. These ideas resonate, however, with long-standing, increasingly global, and often taken-for-granted ideas about the desirability of rational governance and superiority of science as an arbiter of contentious disagreements.

The interconnected processes of neoliberalization and scientization are visible in Mexican and Canadian agricultural and environmental policies. In both countries, debates about biotechnology have been dominated by neoliberal policy ideas and a scientized approach to environmental governance. These policies "normalize" the development and commercialization of GE crops, to use the terminology of Jasanoff (2005, 95), even as challengers take issue with the new technologies. In both Mexico and Canada, policies toward biotechnology marginalize social, economic, and ethical concerns, leaving farmers and activists with few official venues in which to express their wide-ranging critiques of the technology and its impacts.

Mexico and the Politics of Maize

Mexico was among the first countries to produce GE crops commercially, beginning with insect-resistant transgenic cotton in 1996. Today, Mexican farmers produce GE soybeans and cotton on over seventy thousand hectares (James 2010a). There has been little public opposition to these GE crops. Nevertheless, plans to introduce GE maize as a commercial crop have been met with enormous resistance from farmers, environmentalists, and scientists concerned about biosafety.

Maize is highly important to the Mexican economy, diet, and culture. More than two-thirds of the gross value of Mexico's agricultural production is in corn cultivation (Henriques and Patel 2003, 24). It is estimated that eighteen million people are either engaged in or dependent on corn production, including at least three million maize farmers (8 percent of Mexico's population) and their families (ibid.). Maize is the basis of the Mexican diet, and also holds symbolic and spiritual significance for many indigenous peoples in Mexico (Esteva and Marielle 2003; González 2001). Additionally, Mexican maize landraces (locally adapted varieties) are globally valued as a source of agricultural genetic diversity. Mexico is the place where maize was first domesticated, about nine thousand years ago, and today is a site of extraordinary maize genetic diversity, thanks to the agricultural practices of small-scale farmers. Although it is difficult to determine precisely how many maize producers use traditional farming methods, maize experts Mauricio Bellon and Julien Berthaud (2006) estimate that more than two million farmers, planting about 5.8 million hectares (80 percent of land used to cultivate maize), are engaged in traditional maize systems. Traditional practices include planting multiple maize varieties to suit different agronomic and culinary needs, saving seed from one season to the next, trading seeds with other farmers and communities, mixing seed of different origins, and adapting improved varieties to local conditions (ibid.). The commercial cultivation of GE maize has so far been prohibited in Mexico because of concerns about maintaining the genetic diversity that these practices generate.

The value of Mexico's maize diversity appears to be almost universally recognized. As Daniela Soleri, David Cleveland, and Flavio Aragon Cuevas (2006, 503) recently put it, "Most people, regardless of their position on GE crops, agree about the importance of diversity." Plant breeders have long valued "centers of origin" like Mexico as the places to find plants with useful traits, such as disease resistance or hardiness in difficult weather conditions. For this reason, plant breeders, ecologists, and environmentalists—not to mention the small-scale producers who rely on local corn varieties for food—are concerned about what may happen to farmers' varieties of maize once transgenes become incorporated into the genome.

Despite the great importance of local varieties, Mexico is now enormously dependent on inexpensive maize imports from the United States. And because GE varieties are so widely cultivated in the United States, they are also entering Mexico's food supply. Transgenic corn shipped from the United States not only becomes food but also may be used as seed. There are studies showing that the grains are planted to produce maize crops as well. In one recent study in four Oaxacan communities, 23 percent of the farmers who obtained maize to eat that they did not grow themselves had also planted some of the grains (Soleri, Cleveland, and Cuevas 2006, 505). Mexican farmers have a history of incorporating "improved" or "modern" seed varieties into their local varieties, in a process of crossbreeding often called creolization—meaning that the progeny of these crosses become farmers' varieties, or landraces. Bellon and Berthaud (2006, 6) note that this may be done every two or three years to improve a population, when farmers perceive that their seed "gets tired" and needs to be revitalized with an outside source of seeds. Maize is an open-pollinating and highly outcrossing plant species, and its pollen can travel distances of between two hundred and three hundred meters (Alvarez-Buylla 2004, 4). Small-scale Mexican farmers usually own several small plots scattered throughout an area. There is rarely much distance between fields, and farmers cannot prevent the exchange of pollen between them (Bellon and Berthaud 2006, 5). These characteristics of traditional maize systems all make the incorporation of GE maize (whether intentional or not) highly likely.

In 2000, two US scientists working with a small Mexican NGO found GE maize growing in isolated rural areas of Mexico, despite the fact that the crops had not been approved for cultivation in that country (Quist and Chapela 2001). Because Mexico is a center of genetic diversity for maize and its wild relatives, the discussion about this discovery immediately turned to environmental safety issues. Yet urban and rural activists pointed to the broader implications, tracing the root cause of the problem to Mexico's trade policies and agricultural reforms, which have resulted in increasing food dependency on the United States. Among most critics, the transgenic contamination of farmers' varieties of corn is understood as an extension of the inequitable trade relationship with the United States along with the resulting loss of community and culture. A small but vocal group of indigenous rights activists, furthermore, argue that the genetic contamination of native varieties of maize constitutes an attack on traditional cultures and indigenous communities' autonomy. As one grassroots development activist in Oaxaca remarked, "There is a broad attack against maize. The matter of transgenic contamination is only one of the forms in which maize is being threatened. . . . [Because of poverty] the *campesinos* [peasants] are working in the United States and so less maize is being produced. . . . So all of these things are together and we see them as an aggression against indigenous people" (March 6, 2006).

These critiques point to deeper problems in Mexican agriculture. The history of state interventions into the Mexican agricultural economy is complex, fluctuating dramatically under each different presidency (see, for example, Barry 1995; Alcántara 1994). From the 1930s through the 1970s, the Mexican government pursued policies to promote food self-sufficiency—largely with success. Corn production was historically subsidized in order to ensure an inexpensive food supply for low-income consumers, but rural poverty remained (and remains) severe. At the start of the 1980s, the Mexican government embarked on an ambitious effort to improve the lot of peasant farmers and food distribution to the poor. The program, called the Mexican Food System, was financed with petrodollars and foreign loans. It provided agricultural subsidies, credit, and crop insurance, and aimed to raise the incomes of peasant producers. The program resulted in national food self-sufficiency in 1981–1982 (Fox 1993). The sharp drop in oil prices in 1981 and resultant debt crisis, though, brought a quick end to the effective, but expensive Mexican Food System. Facing the debt crisis of the early 1980s, Mexican leaders decided to radically transform and strip down government interventions into agricultural markets, thereby ending the pursuit of food self-sufficiency.

The neoliberalization of Mexican agricultural policy has been widely analyzed and discussed (see, for example, Bartra 2003; Lind and Barham 2004; Preibisch, Herrejon, and Wiggins 2002; Henriques and Patel 2003; Ochoa 2000; Barkin 1987; Brown 2004; Fitting 2006a; Henriques and Patel 2004; Alcántara 1994). The first steps of this transformation occurred in the mid-1980s, under the presidency of Miguel de la Madrid, who negotiated Mexico's 1986 accession into GATT, which ultimately became the WTO. The de la Madrid administration dramatically reduced subsidies to agricultural producers. The subsequent presidency of Carlos Salinas de Gortari, from 1988 to 1994, further solidified the commitment to neoliberalization. Salinas de Gortari negotiated Mexico's participation in NAFTA with Canada and the United States. Neoliberal reforms affected access to land and seeds as well as subsidies for maize production.

In 1992, a significant reform to Article 27 of the Mexican Constitution ended the government's commitment to land redistribution through the provision of state-owned lands that were set aside for the use of farming communities, called *ejidos*. The reforms enabled *ejiditarios* (residents of ejidos) to sell parcels of land to private owners (Harvey 1998; Cornelius and Myhre 1998). This reform marked a radical break from the historical role of the Mexican state in supporting peasant agriculture. There have also been dramatic changes in Mexico's stance on plant breeding and genetic resources. The Mexican Federal Law on Plant Varieties, passed in 1996, was a decisive turn away from public plant breeding in favor of a commercial, for-profit system of seed development. Previously, the 1961 National Seed Law had given the government exclusive control over plant-breeding activities, treating the development of improved seed as a public good and prohibiting private research into seed improvement without approval from the secretary of agriculture (Martinez Gómez and Torres 2001). In the 1960s and 1970s, Mexico had a prominent role in international debates about plant genetic resources, pushing for the creation of a "non-commodified system for the use and conservation of agricultural germplasm" (ibid., 289). After the 1980s, however, there was a "change of the guard" in the agriculture department, and public spending on seed development was reduced, while private firms rapidly entered the seed market (ibid.). Subsequently, as a condition of participation in GATT and membership in NAFTA, Mexico had to adopt national legislation to create intellectual property rights for seed developers.

Under pressure from the World Bank, the Salinas de Gortari administration also reduced its subsidization of corn production, transforming the strategy from self-sufficiency to import dependency (Barry 1995). Mexico illustrates the implications of transforming agricultural economies along the lines of so-called comparative advantages. It can produce fresh fruits and vegetables year-round, with low labor costs, while the United States can produce corn in large quantities at low prices. When NAFTA was negotiated, the understanding was that with the elimination of subsidies for corn production, a "significant proportion of displaced [corn] producers would enter the national labor market and would be employed in sectors of higher productivity. The overall outcome would be a positive contribution to the trade balance and to Mexico's fiscal accounts" (Nadal 2002, 1). From the beginning of NAFTA in 1994 until 2002, Mexico's exports of fruits and vegetables increased 57 percent (Henriques and Patel 2003). At the same time, as expected, imports of corn doubled from three to six million metric tons (Nadal 2003). Indeed, the Mexican government went beyond the terms set out in NAFTA for the transition away from corn production, phasing out tariffs on corn imports in just thirty months, rather than the agreed-on fifteen years (Nadal 2003). Cheap US corn flooded the market. In addition to introducing transgenic seeds, the major effect of increasing dependence on imported corn was a significant reduction in corn prices. This contributed to the sharp increase in rural poverty along with a corresponding growth in migration from the countryside to the cities and the United States (Fitting 2006a). Because of these changes, farmers have experienced dramatic losses, including the loss of community and culture when people leave the land to seek their livelihoods elsewhere.

Some studies, however, suggest that confounding all expectations, Mexican maize production has not decreased. In fact, the area in maize production appears to have actually increased in the poorest Mexican states—Oaxaca, Chiapas, Veracruz, and Guerrero—indicating that the poorest households are expanding subsistence maize production in response to economic vulnerability (Nadal 2003). Maize production, in other words, while not economically profitable, is a survival strategy for the rural poor who do not migrate. It may also suggest, as some scholars have asserted, that some farmers are resisting neoliberalization by increasing their commitment to traditional farming practices (Barkin 2002).

In the midst of these radical changes in Mexican agriculture, GE crops were introduced to the market, and the Mexican government has struggled to construct and implement appropriate regulations. Mexican rules for GE product approvals have been under almost constant transformation and renegotiation since the first applications for experimental trials in 1988 (Alvarez-Morales 1995, 2000). Mexico first set formal rules for the importation and establishment of experimental field trials of GE crops in 1996, and began to actively participate in the deliberations for the UN Biosafety Protocol in June 1997 (Aguirre González 2004, 193). By putting the issue of biosafety on the political agenda, these international negotiations were an important stimulus to the further development of regulations to protect biological diversity (Gupta and Falkner 2006).

Since 1998, Mexico's SEMARNAT has taken an increasingly central role in the politics of biotechnology, providing a modest counterweight to the agriculture agency, Secretaría de Agricultura, Ganadería, Desarrollo Rural, Pesca y Alimentación (SAGARPA), which overtly favors biotechnology (Gupta and Falkner 2006, 35). In 1998, SAGARPA announced a de facto moratorium on the field release of transgenic maize. Yet the moratorium did not account for transgenic grain, which was arriving, unregulated, from the United States (Turrent and Serratos 2004).[7] Concerned about the lack of an adequate regulatory framework, Mexico's scientific community pushed the government to formalize a process of scientific risk assessment for biotechnology. In April 1999, twenty-one Mexican scientists presented a signed statement to the president calling for action on the risk assessment of biotechnology and creation of a scientific agency responsible for biotechnology evaluation (Aguirre González 2004, 196). In response to this pressure, the government created the Interministerial

Commission for Biosafety and Genetically Modified Organisms (CIBIO-GEM), an interagency governmental body that formulates and coordinates policy on the biosafety of GE crops. Still, until 2005, Mexico's legal framework for the regulation of biotechnology and biosafety was "fragmented and dispersed" (Gutiérrez González 2010, 114).

From 1999 to 2005, Mexico's federal elected officials debated the creation of a new Biosafety Law for Genetically Modified Organisms (LBOGM). The Biotechnology Committee of the Mexican Academy of Sciences, led by Francisco Bolívar Zapata, a noted researcher in biotechnology, was particularly influential in the formation of Mexico's new rules for GE crop regulation (Greenpeace Mexico 2005). Other proposals were greatly discussed, but faced tough opposition from the biotechnology industry, which in 1999 formed an organization, Agrobio, to represent its interests in Mexican politics (Aguirre González 2004). NGOs, scientists, and celebrities alike protested the proposed law, saying it would serve only to promote the interests of the biotechnology industry, and did not sufficiently protect biodiversity or the rights of peasant farmers and indigenous people (Massieu Trigo and San Vicente Tello 2006). Massive campesino protests in 2003 led to the signing of the National Accord for the Countryside, which, among many other points related to Mexican agricultural policy, resolved that campesino organizations and producers had to be consulted about the biosafety law (ibid., 44). In 2004, Oaxacan farmers and activists took a stand against GE maize, writing a manifesto that emphasized the social, economic, and cultural importance of maize, called on the Mexican government to stop the importation of transgenic maize, and demanded that the biosafety law support both biological and cultural diversity.[8]

Despite such protests, a version of the LBOGM dubbed by activists the "Monsanto Law," because of its promotional stance on biotechnology, was ratified in February 2005. The law created a process for testing and approving transgenic crops for commercial release. The law called for the creation of a "special regime" for the protection of maize biodiversity. For four years, under much public scrutiny, public officials struggled to establish the special regime for maize. In this process, Mexican officials offered formal assurances that protecting maize biodiversity is a priority, particularly in regions where wild maize varieties are present. Nevertheless, the government placed no limitations on importing transgenic maize from the US. Finally, in March 2009, by presidential decree, Mexico's LBOGM was amended to establish rules replacing the special regime for maize, opening the door to the future commercialization of GE maize seeds. Experimental field trials of GE maize have taken place since then in several northern Mexican states (James 2010a). Until the amendment of the LBOGM in 2009, activists continually and largely successfully fought the initiation of experimental trials of GE maize by pointing to the unfulfilled requirement to set up the special regime. For some critics of GE maize, those short-term victories, based on arguments about environmental risk, offered a "window of opportunity to rethink the role of the countryside and agriculture in our country, based on new principles" such as sustainable agriculture and the value of small-scale production (Nadal 2006). Yet in a context where assertions based on environmental risk are clearly more in keeping with the government's priorities, advocates of such alternatives found limited opportunities for advancing demands for agricultural reform at the federal level. Indeed, regulatory officials have been largely unresponsive to the far-reaching demands of maize activists, which include protections for small-scale agriculture and a withdrawal of staple crops from NAFTA.

One of the ways that the Mexican government has justified its unresponsiveness to public concerns about GE maize is to define the issue as narrowly technical. Cultural and economic issues, while raised persistently by activists and often mentioned by sympathetic government officials when I interviewed them, are not included among established regulatory criteria. The LBOGM established a public participation procedure, in which members of the public may give their opinions within twenty working days of a request for GE crop authorization. But Article 33 of the law specifies that these opinions must be "supported technically and scientifically," thus explicitly ruling out opinions about biotechnology that do not conform to the criteria of rational risk assessment. Thus, policies toward biotechnology marginalize social, economic, and ethical concerns, leaving farmers and activists with few official venues in which to express their wide-ranging critiques of the technology and its impacts.

Canada's Canola Controversies

Canola is the product of a Canadian state-sponsored plant-breeding program during the 1960s and 1970s, which successfully transformed rapeseed into an oilseed suitable for human consumption and named it canola.[9] As a result, the rapeseed industry underwent a dramatic transformation, "changing from a small, niche product produced in a few developing countries to the third largest source of oils for human consumption, produced and used in many developed agricultural nations" (Phillips and

Isaac 2001, 244). Canola today is one of the world's most popular edible oils and one of the major crops produced on the Canadian prairies. Bringing fourteen billion dollars per year to the Canadian economy, canola is enormously important (Canola Council of Canada 2008). Canada is by far the world's top exporter of canola, topped in canola production only by China, which exports only a small percentage of what it produces. According to the UN Food and Agriculture Organization, the second-largest exporter of rapeseed and mustard seed (the category including canola in these statistics) is France, which exports less than half of that exported by Canada. Thus, of all the world's canola producers, Canada is the most export dependent. Canada produced 9.86 million tons of canola in 2004, and exported over 68 percent of the canola it produced (Phillips and Isaac 2001, 247).

For the most part, Canadian farmers have eagerly adopted GE canola since it was first introduced in the mid-1990s, and by 2009, 93 percent of the canola grown in Canada was GE (James 2009). In the late 1990s and early 2000s, though, there was still a sizable proportion of farmers—both conventional and organic—who cultivated non-GE canola. For those farmers, it was practically impossible to avoid contamination, which posed a variety of threats, including accusations of patent infringement (for saving patented GE seeds) and the loss of sales of organic products to countries where GE foods are not accepted. These incidents differ from those in the case of Mexican maize in significant ways. Despite its economic importance, canola does not have the historic, environmental, and cultural significance that maize has in Mexico. Furthermore, transgenic canola has been deregulated in Canada, so when contamination occurs, it is from approved varieties, unlike in Mexico, where GE maize is not approved for commercial cultivation.

Tiny canola seeds are dispersed when they are carried from fields and trucks by the wind or animals, and pollen from canola fields can extend over long distances. The "promiscuity" of canola has led to contamination of the seed supply. Routine testing for transgenes does not take place, yet a number of studies have shown that genetic contamination is widespread (Downey and Beckie n.d.; Friesen, Nelson, and Van Acker 2003). In addition, experimental studies have found evidence of pollen flow ranging from 366 meters in Canada (Stringam and Downey 1982) to 4 kilometers in the United Kingdom (Thompson et al. 1999). In 2002, a team of Australian researchers studying pollen flow in working, commercial canola fields confirmed the experimental data suggesting long-distance pollen flow (Rieger et al. 2002). A small amount of pollen was found to travel up to 3 kilometers from its source, carried by wind and insects. One of the authors of the Australian study commented that because the amount of gene flow at long distances is not large, it should not pose a problem for keeping levels of transgenic material in conventional crops below 1 percent—the level of contamination accepted by Australian regulators (Stokstad 2002). Yet the findings indicated the difficulty, if not impossibility, of completely preventing the hybridization of transgenic and nontransgenic crops once they are released into the environment. For farmers who prefer not to grow GE canola as well as consumers who want their food to be non-GE, this choice is increasingly restricted.

Just as in Mexico, conflicts over GE canola take place in the context of neoliberal agricultural reforms. The shift in Canada's agricultural economy began in the early 1980s, when the government broke from the Keynesian economic policies that were instituted in response to the Great Depression. The agricultural support programs created in the 1930s and 1940s and the post–World War II agricultural boom created a period of relative stability and prosperity for Canadian farmers, although the increasing industrialization of agriculture led to greater concentration and a steady decline in the numbers of family farms. In the 1960s and 1970s, worldwide changes in the agrifood system (Friedmann 1982) began to threaten the economic stability of Canadian agriculture.

In the 1980s, Canadian policymakers started to adopt a radically different approach to agricultural policies, influenced strongly by the emerging neoliberal approach to economics advocated by US president Ronald Reagan and British Prime minister Margaret Thatcher. As Canadian sociologist Murray Knuttila (2003, 300) put it a few years ago:

The past two decades represent a virtual revolution in Canadian agricultural policy. Various Canadian governments, under the ideological umbrella of free-market economics and a more laissez faire approach to politics, have demonstrated a propensity to move away from supporting and encouraging agricultural production. . . . The new "policies" are clearly designed to allow the logic of the market to restructure prairie agriculture.

Many agricultural support programs, previously thought to be politically untouchable, were eliminated. Darrin Qualman and Nettie Wiebe (2002) of the National Farmers' Union (NFU), a membership organization that advocates for policies that support family farms, describe these changes as structural adjustment, noting the multiple ways in which the transformations in Canadian agricultural policy mirror those forced on developing countries by the International Monetary Fund and World Bank during the same period. Key features of these changes include a shift from production for national markets to production for export, dramatic cuts in government spending, deregulation of the grain-handling system, measures to attract foreign investments, the privatization of government industries and utilities, and the removal of subsidies, price controls, and other supports (ibid.).

The shift in emphasis to producing for export markets has increased Canadian farmers' vulnerability, putting them in direct competition with producers around the world. From 1989 to 1996, Canadian farmers doubled their exports from $10 billion to over $20 billion (ibid., 4). Even as exports dramatically increased, since the 1980s, Canadian farmers have faced a perpetual "farm crisis." Net farm income has fallen to levels not seen since the Great Depression. For instance, farmers in the province of Saskatchewan earned an average net income of just $1,783 per farm in 1999 (Qualman 2001, using Statistics Canada data). Many farms realize *negative* net incomes, depending for survival on off-farm labor and crushing debts. The result has been rapid farm loss and massive population declines in rural provinces. Many blame this on the continued use of subsidies by the European Union, which put European farmers at an advantage. But some critics have observed that foreign subsidies are not to blame for low prices for export crops; instead, concentrated transnational agribusinesses push world prices down to their benefit (Qualman 2001). A study of the Canadian agrifood commodity chain, from fuel and fertilizer to grain elevators and food processors, found that only a handful of enormous firms dominate each sector (ibid.). Through mergers and takeovers, the number of firms in competition at each link of the commodity chain is shrinking, and firms are also becoming more vertically integrated (Heffernan 2000). Without significant competition, these firms exert downward pressure on their buying prices and upward pressure on their selling prices.

Neoliberal reforms that privatized plant breeding are among the transformations that encouraged the growth and concentration of large agribusiness corporations. Previous to 1985, the development of new plant varieties for Canadian agriculture was almost entirely the responsibility of public breeding programs at Agriculture Canada and public universities (Kuyek 2004; Phillips 2003, 122). Since the 1980s, however, seed development has shifted primarily to private companies, which have merged into a small number of gigantic transnational corporations. The Canadian government accomplished this transformation not by passively withdrawing from the domain of plant breeding but rather through a series of active interventions, including direct subsidies to private seed companies, new laws and regulations pertaining to intellectual property and seed saving, and policies to dismantle public breeding programs (Kuyek 2004, 2007a, 2007b). Since the early 1980s, patents on gene sequences have been granted in Canada, making it possible for biotechnology companies to demand royalties for the use of GE seeds. One effect of these changes is to severely limit or eliminate seed saving by farmers, who now must purchase fresh seed in order to avoid patent infringement. Monsanto, the world's largest seed dealer, has been particularly aggressive in defending its patents, frequently pursuing legal action against farmers suspected of saving seeds.

Public opposition to GE crops in Canada has therefore taken shape in the context of significant tensions in the agricultural sector. The NFU has been highly critical of both neoliberal shifts in Canadian agricultural policy and the growing use of agricultural biotechnology. In a 2000 policy statement, the NFU pointed to the ways that GE crops deepen the economic difficulties for Canadian farmers. The statement noted that "fertilizers, chemicals, and other technologies failed to fulfill their promises of farm profitability, [so] many farmers rightly question the economic benefits of genetically modifying crops and livestock." Furthermore, the NFU indicated that important markets are rejecting GE food, and in particular, "the proliferation of some GM crops has effectively deprived many organic farmers of the option to grow those crops." In regard to this, the NFU was explicit: "Genetic pollution seriously erodes the incomes of organic farmers and those who do not use GM seeds. Government must hold genetic modification companies accountable for the costs their products create for other farmers and the general public." Finally, referring to the intellectual property rights that biotechnology companies claim, the NFU criticized GE crops because they "give corporations increased control over family farms" (National Farmers Union 2000).

Regulatory policy for agricultural biotechnology in Canada, however, has not addressed the economic concerns that critics like the NFU raise. Instead, the regulation of GE crops relies on a science-based risk assessment. One analyst of this system, policy scholar Elisabeth Abergele (2007, 174), remarks that Canadian biotechnology policy reflects the country's export orientation:

Canadian governance of agriculture and more specifically agricultural biotechnology are largely structured around the country's ability to open agricultural export markets. . . . [T]he principles used to regulate the commercial release of GE crops were derived for this purpose, placing emphasis on the design and implementation of scientific rules and standards leading to predictable biotechnology governance nationally and internationally.

Since the early 1990s, Canada has been an active player in international deliberations, through the Organization for Economic Cooperation and Development, the WTO, the Biosafety Protocol, and other institutions, working to define the terms of global biotechnology governance. Regulatory officials from the Canadian Food Inspection Agency (CFIA) routinely participate in negotiations regarding international scientific and technical standards, which aim to harmonize international regulatory frameworks to facilitate trade (Abergel 2007). Canada's position in these negotiations and its own regulatory procedures is that biotechnology, as a process, poses no special risk. Rather, each "novel" product should undergo a scientific risk assessment. Often working with the United States on international trade matters, Canada has been a vigorous promoter of the notion of the science-based regulation of GE crops.

Within Canada, the 1993 Regulatory Framework for Biotechnology made three existing agencies responsible for regulating GE crops and foods: the CFIA, Health Canada, and Environment Canada. The CFIA and Health Canada, the two agencies that handle most of the approvals, both use the concept of "substantial equivalence" as a principle for carrying out assessments. This means that using applicant-supplied data, a GE plant will be compared to an unmodified counterpart to determine whether they are similarly "safe." For example, the CFIA compares along five dimensions: "altered weediness potential; gene flow to related species; altered plant pest potential; potential impact on non-target organisms; and potential effect on biodiversity" (Andrée 2002). If the GE plant is determined to be substantially equivalent to its conventional counterpart or the risks can be mitigated, the plant is approved for commercialization. No follow-up or long-term studies are carried out by the government. Consumers, environmentalists, and other citizens generally do not have a say in the product evaluations (ibid.), and there are no requirements for labeling transgenic products in Canada. In other words, once Canadian regulatory scientists have determined that the crops are substantially equivalent to conventional crops—along a narrow range of criteria—they are not treated differently than any other crop plant.[10]

In 2000–2001, the Expert Panel of the Royal Society of Canada (2001) undertook a detailed evaluation of Canada's regulatory system for biotechnology, concluding that the principle of substantial equivalence was deeply flawed, and that the system as a whole lacked both transparency and scientific rigor. This finding is worrisome because the regulatory process excludes public participation in the name of science-based decision making, but the science itself, according to the Royal Society's review, is flawed. This concern is echoed by other critics, who point out that regulators often know very little about the environmental effects of the "baseline" crop varieties to which they are comparing the novel crops (Andrée 2002). When the report was released, prominent plant researcher R. Keith Downey (2001) publicly criticized the report as being "biased to the extreme" against biotechnology, but critics of GE crops welcomed the report and called for the implementation of its numerous recommendations. The Canadian government created an "action plan" to address some of the concerns raised in the Royal Society report, but a 2004 study indicated it had yet to realize most of the recommended changes (Andrée and Sharratt 2004).

In addition to the central concern with the scientific evaluation of the safety of biotechnology, the Expert Panel of the Royal Society of Canada (2001, 223) report also recognized that there were "broader social, political and ethical considerations" relevant to the regulation of GE crops, particularly in the labeling of food products. Activist groups supported this recommendation. For example, the Polaris Institute, which works to support citizen movements in Canada, echoed the Royal Society report's recognition that decisions about biotechnology "should not . . . be based only on independently verified experimental data related to health and environmental risks, but also on an examination of socio-economic issues and ethical concerns" (Andrée and Sharratt 2004, vii). In 2010, in the wake of the costly discovery that Canadian flax exports to Europe were contaminated with a discontinued variety of GE flax, the Canadian Parliament debated a piece of legislation that would have required an assessment of potential market losses before approving the commercialization of GE crops. Had it been approved, the legislation would have been a departure from the science-based approach that has dominated the regulation of GE crops since the 1990s. The legislation did not pass, though, and the regulatory process remains limited to a constrained set of criteria for the determination of safety.

In summary, the Canadian government has been a strong supporter of the biotechnology industry and has constructed a regulatory system that is aligned with broader commitments to globalized agricultural trade. While certain GE crops, such as canola, have been widely adopted, it has been at the expense of those farmers who choose to avoid GE seeds—whether because they practice organic agriculture, wish to continue saving their own seeds, or want to sell their products in export markets that reject GE foods. Rather than responding to multifaceted public concerns (both domestic and international) about GE crops and the changing structure of the agrifood system, Canada has advocated science-based risk assessment as a shared regulatory framework around the world.

Constraints and Opportunities

I have argued in this chapter that, during a period of neoliberalization that has transformed agrifood systems worldwide, new forms of regulation have been created to address public concerns about biotechnology. Specifically, there is an increasingly vehement emphasis on "science-based" assessment of technological developments and regulations to manage objectively defined risks. Like neoliberal theory, the concept of science-based governance defines areas into which the state should not venture. Namely, governments should not hinder the flow of trade or the exertion of private property rights without strong scientific proof to support the need for intervention. The economic, cultural, and ethical consequences of technological change are marked as "political," and therefore unsuitable, reasons to obstruct the commercialization of GE seeds. In both Mexico and Canada, rules governing transgenic crops have thus failed to address public criticism of the systems of agriculture that they facilitate. As opposed to grappling openly with the implications of agricultural biotechnology for agricultural industries and rural livelihoods, federal regulatory agencies in both countries have focused narrowly on human health effects and some environmental issues.

Certain features of this political terrain have generated opportunities for activists to resist the expansion of GE crops and protect alternative systems of agricultural production. As a signatory to the Convention on Biological Diversity (CBD) and Biosafety Protocol, Mexico has some obligations to protect native maize varieties and their wild relatives. Observers have noted strategic opportunities for using these international obligations in order to make headway in addressing the deeper "state of emergency for Mexican maize" (Wise 2007). Activists worldwide have referred to Mexico's international environmental obligations when pressing the Mexican government to take action against transgene flow. Moreover, Mexico is one of three members of NAFTA, which liberalizes trade with the United States and Canada. Although NAFTA is in large part the reason for the flood of US-produced maize into Mexico, the treaty also has an environmental "side agreement" that has provided a crucial venue for activists concerned about transgene flow.

Efforts to defend the production of non-GE canola have been less successful than the struggle against GE maize in Mexico, suggesting that without significant support from the international scientific community and the application of biodiversity protections, farmers and anti-biotech groups lack sufficient leverage to block the commercialization of GE crops. Still, activists directly challenge the biotechnology industry through a variety of strategies, including mobilization through the legal system and pressure on companies to withdraw GE seeds from the market. In particular, Canada's dependence on export markets offers a different kind of opportunity to resist GE crops. Consumer opposition in important export markets provides leverage to farmers and activists in Canada. The regulatory system itself, furthermore, with its stated goal of science-based decisions, has become vulnerable to attack on the grounds of faulty science. Farmers have taken advantage of opportunities in the judicial system to call on counterexperts who question the government's safety assessments.

Neoliberal agricultural reforms and the scientization of biotechnology politics, in sum, have constrained public debate about the social implications of agricultural biotechnology; yet among these constraints, antibiotech activists have found some opportunities to resist the further expansion of GE crops. Opponents of GE crops frequently appear to accept the dominant terms of debate, calling for better risk assessment and demanding consumer choice in the marketplace. Critics of GE crops raise serious concerns about how GE crops might affect ecosystems, agricultural systems, and human and animal health, and question whether their government is doing an adequate job of protecting them. At the same time, the politics of genes out of place involve more than just scientific questions about the safety of the new crops, and expansive social-change goals are often contained within arguments about risks. Critics of biotechnology usually view GE crops as incompatible with the kinds of agricultural systems and rural communities that they are trying to build. These aims along with the strategies used in pursuing them are examined in detail in each of the following chapters.

3

The Maize Movement and Expert Advice

In March 2004, environmentalists and maize producers converged on the Mexican city of Oaxaca, holding a protest at a public symposium in which a panel of esteemed scientists was to address the topic of GE maize and biodiversity. The demonstrators crowded into the conference room and proceeded to take control of the microphone, presenting hours of testimony. The activists voiced their opposition to GE maize through protest signs and theatrical interventions, such as the placement of colorful mosaics of maize on the floor. While, as one advisory group member put it, some of the speakers seemed to be "crazed Americans who probably chewed too much peyote," there were many statements from peasants and indigenous people, some dressed in traditional clothing, who spoke about their own experiences, traditions, and beliefs about maize to the expert panel (US scientist, telephone interview, May 30, 2007). To assert the relevance and credibility of their perspective, farmers and other activists presented themselves as the voice of the group most directly and profoundly affected by GE maize: the "people of maize."

Although activists believed the forum was rightfully a space for protest, the experts who participated on the panel held a different view. Expecting reasoned discussion of the background papers that had been prepared, many were dismayed to find seemingly chaotic protests and testimony given at length by the farmers and activists in attendance. As one ecologist described it to me, "They [took] over the meeting and made long speeches, and we listened, and they gave us tortillas and lit candles and did all kinds of things [laughing slightly], and so it turned into more of a protest than a dialogue, and I would have really liked to have more dialogue" (US scientist, telephone interview, September 8, 2005). Another scientist recalled the experience as "torture" (Mexican scientist, Mexico City, November 11, 2005). Still another portrayed his experience this way:

[It was] a really politically charged atmosphere, with banners—one I will never forget said "Death to [Secretary of Agriculture] Victor Villalobos, the king of transgenic crops." . . . Dr. Sarukhan [the committee chair] said to me, "Why aren't you saying anything," [and it was because] I'm not a fool. . . . I wasn't going to put myself at risk. [I]t was a really aggressive environment and of course completely dominated by the opposition groups. (Mexican scientist, Mexico City, November 14, 2005)

The people holding those signs and giving testimony were participants in what some were then beginning to call a "movement in defense of maize"—a mobilization of environmentalists, agroecology promoters, indigenous communities, and maize producers, which I will refer to as the "maize movement" or "maize activism." This chapter focuses on one crucial episode in the early phases of this movement. In 2002, a group of activists, scholars, and rural community leaders composed a petition demanding an assessment of the impacts of transgenic introgression into landraces (farmers' varieties) of maize in Mexico. The petition was sent to the Commission for Environmental Cooperation (CEC), a trinational body created as part of the environmental side agreement to NAFTA. The CEC responded, and after many months of research and deliberation, the scientific experts who were selected to investigate the topic held a public meeting in the heart of the state where native maize contamination was first discovered. At the symposium, the ecologists, molecular biologists, social scientists, and other experts serving on the panel were surprised to encounter an impassioned, lively, and (to their eyes) chaotic public protest. Hundreds of farmers and environmental protesters used the symposium as a venue to reframe the maize question in terms of cultural values, the consequences of globalization, and indigenous peoples' autonomy.

The meeting in Oaxaca forged a new relationship between the movement and scientific expertise, blurring the boundaries that typically (if artificially) separated technical assessment from political contention. When the expert advisory group finally released its recommendations several months later, the protesters' influence was unmistakable: the ostensibly scientific report addressed themes of democratic participation, cultural values, spirituality, and sustainable livelihoods. Remarkably, scientific experts in this case became important movement allies, conveying many of the activists' concerns in their final document. This case is not only a key episode in the antibiotech movement, in which genes out of place visibly became a social problem. It also offers critical insights on the relationship between science and social movements. As a growing number of social movements are engaged with issues informed or even created by new developments in science and technology, how are scientists drawn in as allies, and how does this change the arguments that scientists make? In this chapter, I suggest that the Mexican maize struggle illustrates a process that I call an "epistemic boomerang," through which local groups and NGOs, frustrated by exclusion from policymaking debates, go outside normal political channels to appeal to scientists, hoping to mobilize scientific research in support of their social goals.

I borrow the metaphor of a boomerang from political scientists Margaret Keck and Kathryn Sikkink (1998). In Keck and Sikkink's boomerang model of transnational advocacy, when states are unresponsive to the demands of their citizens, activists may seek the support of international allies in a process referred to as externalization (Tarrow 2005). The boomerang metaphor, further elaborated in a variety of recent studies (Tarrow 2005; Seidman 2007), is meant to capture the sense in which local citizens voice their concerns to the international community, and those concerns are then echoed back as external pressure on the citizens' recalcitrant governments. Pressure comes from governments and publics that are sympathetic to local groups, or from international agencies that have the authority to make decisions or recommendations about domestic issues. This boomerang pattern can be an effective way to produce political change when governments are deaf to the grievances of their own citizens.

Externalization can take the form of information politics, in which activist networks diffuse information to international allies about abuse or injustice taking place locally, or institutional politics, in which local actors take their grievances to supranational institutions, such as the European Court of Justice or the United Nations (Tarrow 2005). Transnational networks of activists and NGOs play an important role in the externalization of contentious politics, often aiding domestic social movements when the efforts of local organizations to pressure their government are rebuffed. I suggest that expert advisory groups and other sources of scientific advice to government exert a kind of influence that may serve as leverage for local social movements. Their influence mainly works through the political authority of science and deference to experts; therefore, I use the term epistemic boomerang.

There is a significant difference between the expert advisers in the epistemic boomerang model and the sympathetic international actors who are frequently found in other types of struggles, such as human rights advocacy. Scientific experts who openly venture into the realm of politics, moving from advice to advocacy, from causal to principled claims, risk their credibility as objective observers (Kinchy 2006; Kinchy and

Kleinman 2003; Moore 1996; Frickel 2004b). In cultural contexts where scientists are seen as biased if they express political opinions, scientific experts are far less likely to explicitly state their political affinity with activist groups. Thus, in the epistemic boomerang model, the key feature is interaction between activists and expert advisers, resulting in scientific advice that furthers the aims of a local struggle. Local groups use their relationships with NGOs to gain access to scientific researchers and advisers, in the hope that the experts will influence state policy by providing advice and making other expressions of scientific concern about the issue at stake. An epistemic boomerang, however, also creates the potential for the two-way sharing of knowledge and ideas between experts and the local activists who are seeking social change.[1]

The formation of an epistemic boomerang depends on scientists' willingness to respond to and interact with social movements. Some organizational and discursive factors may encourage scientists to be responsive to activist concerns. First, activists must frame their concerns in ways that resonate with questions that interest scientists, whether in academia, research organizations, or the expert branches of government.[2] The organizational setting for interactions between activists and experts is another significant factor in the formation of an epistemic boomerang. In particular, the site for interaction must be public and accessible, either through the organization of physical space (such as public hearings) or through policies encouraging citizen input and participation. UN meetings, for example, have increasingly opened up to NGOs. In such a context, experts do not simply give advice to activists or political authorities; they also listen to local groups seeking political change. Could this lead experts involved in an epistemic boomerang to make principled claims on behalf of a social movement? Under what circumstances do scientists break with convention and take moral positions in support of local activists? I explore these questions in the case of the mobilization surrounding the CEC.

The Maize Movement

Mexican environmental activists started to address the issue of GE crops in the late 1990s, beginning with a campaign against the importation of GE maize from the United States. The tactics were influenced by activism occurring worldwide. In 1996, Greenpeace International initiated a major new campaign related to GE crops, starting with a blockade of the ports where Monsanto's GE soybean seeds were entering Europe, unlabeled

(Schurman and Munro 2006, 25). The tactic of targeting grain shipments was replicated in Mexico in early 1999. Mexican Greenpeace activists took samples gathered from ships bringing corn from the United States to Mexico and sent them to a government laboratory in Vienna. The corn shipments were found to contain Bt varieties—the type of genetically modified corn that produces its own pesticide. These findings, viewed by Greenpeace (1999) as a form of contamination, were announced in the press and made public through Greenpeace demonstrations at the port of Veracruz. While the action in Veracruz did not spark the kind of consumer response seen across western Europe, it marked the beginning of a contentious struggle over transgenes in Mexican maize. Other organizations also started investigating the implications of biotechnology. In 2000, one environmental NGO based in Mexico City published a report on the status of GE crops in Mexico, arguing that the introduction of GE maize threatens to "homogenize the enormous variety of Mexican maizes. The effect of this is not only a culinary problem. The loss of biodiversity is perhaps one of the most grave dangers that confronts the planet today," reducing the possibility of living sustainably on the earth. Furthermore, the report maintained, the introduction of GE crops threatens to replace cultural diversity with industrialized seeds and crops, curtailing "the freedom to decide and choose what to eat [and] the freedom of *campesinos* to grow food organically and to choose and save their seeds for the next season" (Gómez Alarcón 2000, 15–17).

In this context, Ignacio Chapela, a scientist at the University of California at Berkeley, decided to help one rural Mexican community test its maize for transgenic material. Chapela had a long-standing relationship with a small development NGO in the state of Oaxaca.[3] In 2000, he sent his graduate student, David Quist, to Ixtlán, Oaxaca, to teach local residents how to conduct polymerase chain reaction (PCR) analyses—a method of detecting transgenic sequences of DNA. The intention was to provide the community with a means of certifying its own maize as non-GE. According to Quist (telephone interview, 2005), he and Lilia Pérez, a local agronomist, conducted a few tests the evening prior to running the workshop. To their surprise, they discovered that transgenic DNA sequences were already present in samples of native corn from rural Oaxaca (Quist and Chapela 2001). Wishing to avoid provoking a panic, Quist quietly brought the samples back to Berkeley for further analysis (Delborne 2005, 155). Quist and Chapela confirmed the findings through further testing, and submitted the results for publication in *Nature*. In the meantime, Chapela shared the findings confidentially with regulatory officials in Mexico. In early September 2001, the officials made the findings public and later announced their own research findings that confirmed the presence of transgenes in maize growing in Oaxaca. The announcement set off a storm of protest among activists and scientists alike, and by the time the story was reported in *Nature* (Dalton 2001) and the *New York Times* (Yoon 2001) in late September and early October, the scandal had already reached international proportions.

The research article appeared in *Nature* on November 29, 2001. The report claimed to have found signs of transgenes in samples of native maize taken from the Sierra Juarez region of Oaxaca (Quist and Chapela 2001). A Greenpeace México e-bulletin on the day that the study was released indicates how much the activists valued scientific reports in this dispute. The announcement pointed out that the previous month, the secretary of agriculture, Javier Usabiaga, said that there was no scientific evidence of maize contamination. Greenpeace México (2001; my translation) responded, "Now we ask him if he thinks this article, published in one of the most important scientific journals in the world, seems like enough evidence to him."

A coalition of NGOs and farmer organizations composed a "public denouncement" (*denuncia popular*)—a formal legal complaint submitted to the Procuraduría Federal de Protección al Ambiente (PROFEPA), the section of the environment ministry in charge of applying environmental legislation. The nineteen-page document compiled all known scientific data, and outlined in detail each step in the discovery and announcement of transgenic contamination in Mexican maize. On the basis of that factual information and Mexican environmental law, the authors criticized PROFEPA for its failure to act to prevent the contamination and called for a ban on the import of GE corn from the United States. The public denouncement was submitted to PROFEPA on December 6, 2001, barely a week after Quist and Chapela's research findings were officially published in *Nature*.

Almost immediately, however, critics attacked the credibility of the Quist and Chapela (2001) study. A variety of scientists submitted technical critiques of the study to *Nature*—two of which were published in April 2002. As Jason Delborne (2008) describes in a study of the controversy, the researchers' methodology and analysis were also attacked in an editorial in *Transgenic Research*. Furthermore, nearly one hundred probiotech scientists tried to diminish any alarm raised by the study by signing a letter asserting that the kind of gene flow Quist and Chapela claimed to find was harmless, and even welcome. Finally, in an unprecedented act, *Nature* published a statement withdrawing support from the Quist and Chapela (2001) article that the journal had originally published (Delborne 2008).

With the evidence of maize contamination in question, maize activists needed further scientific studies to support their positions. Yet such studies remained "undone" (Woodhouse et al. 2002; Frickel et al. 2010). From the time that Chapela informed the Mexican government that he had discovered transgenes in maize growing in Oaxaca, Sol Ortiz-García of Mexico's National Institute of Ecology (a government agency) had been monitoring the presence of transgenes in that region. In 2002, Ortiz-García and her colleagues announced that they possessed data to confirm Quist and Chapela's findings of low levels of transgenes in samples of native maize (Ezcurra and Soberón Mainero 2002). Their report was not accepted for publication in *Nature*, though, and the results were never published in any other peer-reviewed journal. Thus, the finding that genetic contamination of native maize had occurred remained officially unreplicated, while the controversy continued through 2002 and 2003.

Even as the scientific debate raged on about whether transgenes had found their way into Mexican maize, and whether it mattered to biodiversity if they had, a variety of advocacy organizations began to articulate a more explicitly political critique of GE maize. Anthropologist Elizabeth Fitting (2011, 116) writes that maize activists in Mexico "[draw] upon transnational movements and the Mexican counter-tradition of valuing peasant knowledge to advocate for maize sovereignty and quality in terms of its taste, its role in *in situ* [in the field] biodiversity, and the quality of the producer's livelihood." That is, activists defend small-scale maize cultivation—using locally adapted varieties and rejecting GE maize—because of the food quality it produces, the contributions it makes to agricultural biodiversity conservation, and the economic benefits for campesinos. This multidimensional understanding of the "maize issue" did not emerge spontaneously but instead was produced in the process of forming an activist network—which itself was "the outcome of decades of rural organizing and farmer-scientist exchanges" (McAfee 2008, 157).

The emerging maize movement relied on coordination and information sharing among organizations based in Mexico City, other urban centers, and numerous rural communities. Organizations previously involved in environmental politics, genetic resources activism, indigenous rights movements, agroecology, and opposition to NAFTA and other neoliberal reforms each critiqued GE maize, and interpreted the reports of contamination from their own perspectives. The Action Group on Erosion, Technology, and Concentration (ETC Group) and Greenpeace, both transnational NGOs, played important roles in organizing events, publishing materials, and sharing information among multiple community groups and organizations in the emerging network in defense of maize.

ETC Group, headquartered in Ottawa, Canada, originated out of the work of Pat Roy Mooney, Cary Fowler, and Hope Shand, pioneering researchers and advocates on issues affecting the preservation and control of agricultural genetic diversity in the late 1970s and 1980s. The three formed an organization called the Rural Advancement Foundation International (RAFI) to study and advocate for issues involving biodiversity, intellectual property, and indigenous agricultural knowledge. RAFI's concerns about plant genetic resources focused simultaneously on privatization and "genetic erosion," both of which were viewed as a threat to the world's food supply.[4] Books like *Seeds of the Earth* by Mooney (1979) drew international attention to the problem of genetic erosion, which occurs when diverse farmers' varieties are replaced by genetically homogeneous "modern" crops. Mooney, among others, argued that the privatization and standardization of seeds leads to the reduction of plant genetic diversity, which in turn could lead to disastrous vulnerability to plant diseases. In 2001, RAFI changed its name to ETC Group. The organization has ten staff members, and offices in Canada, the United States, Mexico, and the Philippines. The office in Mexico City opened in 1999, with two staff members (ETC Group n.d.).

Greenpeace is a far larger organization than ETC Group. Begun in 1971 in order to "bear witness" to the ecological impacts of underground nuclear testing, Greenpeace now has offices in over forty countries and campaigns on a wide range of environmental issues. Greenpeace is well known for its controversial direct actions, but is also frequently a presence at UN negotiations and in national policy debates. A Greenpeace office opened in Mexico City in 1993, with organizers first protesting against the importation of toxic waste to Mexico, and later carrying out campaigns to protect whales and forests as well as address problems of pollution and waste disposal (Greenpeace México 2011). The office has a small staff and counts on numerous volunteers to participate in visually striking demonstrations. In 1999, Greenpeace México kicked off a campaign against GE food, focusing on the maize that was being imported from the United States.

As the maize scandal unfolded, Greenpeace and ETC Group formed alliances with rural communities through other civil society organizations, Mexican farmers' unions, and small NGOs promoting agroecology. For example, the Unión Nacional de Organizaciones Regionales Campesinas

Autónomas (UNORCA, or the National Union of Regional Autonomous Campesino Organizations) has played a key role in protests against GE maize. UNORCA advocates on behalf of small-scale farmers and was a central organization in the El Campo no Aguanta Más (The Countryside Can't Take Any More) demonstrations of December 2002 and January 2003. In those protests, tens of thousands of peasants and farmworkers marched in Mexico City, and at one point created the largest assembly of peasants there since the 1930s. The protests led to the signing of an agreement between campesino groups and the government, called the National Rural Accord—a compromise that provoked ruptures among the peasant organizations involved in the movement. Organizations, like UNORCA, that were unsatisfied with the agreement, continue to fight against neo-liberal agricultural policies, demanding a renegotiation of NAFTA and organizing demonstrations against the WTO (Sanchez Albarrán 2004; Puricelli 2010).

UNORCA is a member of the Via Campesina (Peasant Way) network, "a global movement that brings together organizations of peasants, small and medium-scale farmers, rural women, farm workers and indigenous agrarian communities in Asia, the Americas, Western Europe and Eastern Europe," and recently, Africa (Desmarais 2002, 94). Via Campesina's membership consists of dozens of organizations from around the globe. It advocates "food sovereignty": "the right of peoples, communities, and countries to define their own agricultural, labor, fishing, food and land policies which are ecologically, socially, economically and culturally appropriate to their unique circumstances" (NSO/CSO Forum for Food Sovereignty 2002). As a member of Via Campesina and a representative of campesinos, UNORCA approaches the matter of maize contamination as a question of rural livelihoods and the loss of farmer autonomy, and frequently joins Greenpeace and other organizations in campaigning against GE crops.[5]

Another key organization in the maize movement is Grupo de Estudios Ambientales (GEA, or Environmental Studies Group), based in Mexico City, which works with rural community organizations to offer hands-on training in agroecological farming practices—that is, agricultural practices that reflect ecological ideas.[6] In Mexico, the agroecology movement draws on the traditional knowledge and practices of indigenous peoples and links the marginalized rural peasantry with contemporary environmentalism (Carruthers 1997). In many respects, GEA's work reflects the values of the Campesino a Campesino (Farmer to Farmer) movement, a decades-long effort across Latin America to facilitate farmer-led approaches to agroecology and support cultural resistance to pressures that drive peasants from the countryside. Campesino a Campesino is "based on principles of agroecology, solidarity, and innovation." It "resists the ecologically degrading and socially destructive commodification of soil, water, and genetic diversity and asserts the rights of smallholders to determine an equitable, sustainable course for agricultural development" (Holt-Gimenez 2006, xvii). The Campesino a Campesino movement, and agroecology principles more generally, emerged in Latin America in the late 1970s as a response to the negative environmental and social consequences of the green revolution—a massive development project aimed at increasing food production through the use of fertilizers, pesticides, herbicides, and hybrid seeds.

Generally speaking, farmers in the Campesino a Campesino movement do not engage in institutional politics. As one promoter of the movement writes, "The people actually producing food and protecting the environment are too busy surviving to engage in institutional debates. . . . For them, sustainable agriculture means sustaining their livelihoods" (ibid., xvi). Nevertheless, farmers who are practicing sustainable agriculture are supported by "hundreds of farmers organizations, nongovernmental organizations (NGOs), development professionals, and individual researchers" that organize the workshops, host the farmer gatherings, and facilitate research that make grassroots sustainable agriculture possible amid the "sea of conventional agriculture" (ibid., xvi, 153).

GEA was founded in 1977 with the aim of assisting rural communities to improve campesino agriculture by incorporating ecological principles and recuperating traditional knowledge. Like many other organizations that promote agroecology in Mexico, GEA was strongly influenced by the teachings of agriculture and ethnobotany professor Efraim Hernández Xolocotzi, who developed the key principles of agroecology based on indigenous practices, such as the milpa system (a mixed field of corn, beans, and squash). Today, among other projects, GEA works with a regional peasant organization in the state of Guerrero to advance local management and natural resources, and develop agroecological projects, including organic fertilization, organic pest control, the conservation and improvement of native seeds, and soil and water conservation. GEA also advocates for public policy in collaboration with Greenpeace and other NGOs. It was among the early critics of GE crops in Mexico, publishing the 2000 report, mentioned above, that addressed the "environmental, political, social, ethical, and health" dimensions of GE organisms (Gómez Alarcón 2000). GEA has played a central role in producing publications and radio programs dealing with the topic of GE crops, is often a cosignatory with Greenpeace on press releases and public statements, and has frequent contact with Mexican public officials.

The organization also plays a self-described "bridging role" between peasant and environmental organizations. This is particularly visible in the Sin Maíz No Hay País (No Country Without Maize) campaign, which adopted the name of a museum exhibition and book that Catherine Marielle of GEA put together with Gustavo Esteva, a Oaxacan scholar-activist (Esteva and Marielle 2003). Continuing the earlier struggle of the El Campo No Aguanta Mas movement, Sin Maiz No Hay Pais activists demand the renegotiation of NAFTA, "to protect our corn, the jobs of millions of farmers, and the way of life in the Mexican countryside"[7] The campaign also calls for a moratorium on GE maize cultivation, the creation of a constitutional right to food, and the passage of legislation promoting food security and sovereignty (Greenpeace México 2007).

UNORCA and GEA are but two examples of organizations that facilitated connections between rural communities and the antibiotech movement in the beginning stages of the Mexican maize movement. Other organizations with critical roles in the mobilization of maize activism in Mexico include Centro de Estudios para el Cambio en el Campo Mexicano (CECCAM, or Center for Studies for Change in Rural Mexico) and Centro Nacional de Ayuda a Misiones Indígenas (CENAMI, or National Center for Assistance to Indigenous Missions). CECCAM is a small NGO that does research on global trade and advocates for peasant agriculture, often collaborating with the organizations previously mentioned. CENAMI is a Catholic organization that provides a wide range of support services to indigenous communities across Mexico, including workshops and projects on agroecology as well as the sustainable use of natural resources. It was crucial in organizing and hosting national meetings of rural organizations on the topic of GE maize. These and other organizations became intermediaries between rural communities and the transnational antibiotech movement.

The indigenous rights movement is an important force shaping the maize movement too. The Zapatista uprising in 1994, timed with the beginning of NAFTA, brought worldwide attention to the struggle of indigenous people in Mexico to claim rights to land, livelihoods, and political autonomy. The Zapatistas provided inspiration to other groups of indigenous people across Mexico, and led to the formation in 1996 of the Congreso Nacional Indígena (CNI, or National Indigenous Congress) to represent at least fifty-six major indigenous ethnic groups in Mexico in a "shared political project" (International Work Group for Indigenous Affairs 1997, 55). The rebellion led to the signing of the San Andrés Accords on Indigenous Rights and Culture between the Ejército Zapatista de Liberación Nacional (EZLN, or Zapatista Army of National Liberation) and the federal government. The San Andrés Accords were supposed to grant autonomy and rights to the indigenous peoples of Mexico plus mark the beginning of a dialogue process, but the Mexican government refused to honor the agreement and continued to repress the Zapatista rebel communities (Collier and Collier 2005; Gilbreth and Otero 2001).[8]

The Mexican maize case was the first instance of transgene flow in a plant that is of great cultural importance to indigenous communities, and indigenous rights organizations have taken positions against GE maize in a variety of forums. Some connections between the anti-biotech movement and indigenous rights groups were already developed previous to the maize scandal, as the result of a high-profile debate about bioprospecting (Barreda 2003). Mexico is home to an enormous variety of genetic diversity, including areas of rain forest. In order to gain access to promising new genetic resources, for use in creating pharmaceuticals or foods, companies since the 1990s have formed bioprospecting contracts that permit them to gain access to indigenous people's knowledge in places across the Global South, usually with some promise of compensation (Hayden 2003). In the 1990s, a controversy arose in reaction to a bioprospecting contract between a community organization in Oaxaca and the biotechnology company Sandoz. Both ETC Group and a neighboring rural community organization accused Sandoz of "biopiracy"—taking biological materials that were the collective resource of the indigenous communities of the region, not a commodity to sell. ETC Group's work in rural Oaxaca laid the basis for mobiling in opposition to transgene flow in maize. More generally, the controversy about bioprospecting in Mexico put the issue of biotechnology and patents on life on the public agenda, garnering a great deal of media attention. Thus, in 2001, when it was reported that transgenes were found in native maize, it was in the context of an already-heated debate about the control held by biotechnology companies over plant genetic resources.[9]

Indigenous activists in the CNI have referred to transgenic maize as one of many threats coming from the liberalization of trade with the United States. In July 2002, the CNI assembly of the center-Pacific region demanded that the federal government "halt the introduction into our country of transgenic maize or maize of doubtful origin" (cited in Vera Herrera 2004). The CNI held the National Forum in Defense of

Traditional Medicine two months later, bringing together representatives of over thirty indigenous peoples from twenty states (Vera Herrera 2004). The public statement made at the forum's conclusion strongly rejected both bioprospecting—a central issue for practitioners of indigenous medicine—and transgenic maize, saying:

As part of our defense of the mother earth and all that is born of her, we repudiate the introduction of transgenic maize in our country, since the mother maize is the primary foundation of our peoples. In consequence, we demand that the federal government declare an unlimited moratorium on the introduction of transgenic maize regardless of its intended use. (cited in ibid.)

As the reference to "mother maize" suggests, for some indigenous communities, maize is not simply a plant but a source of life as well. The introduction of transgenes has been discussed at times as a corruption of the plant's soul, akin to an attack on the health of a family member (González 2006). Indigenous rights activists also made the case that the genetic contamination of native varieties of maize constituted an attack on traditional cultures and indigenous communities' autonomy, as this quotation suggests:

For us maize is sacred, it is present with us [*presente*], and we care for it through all that we do, listening to the wisdom of the elders, respecting our customs and culture. We don't want and we won't allow any transgenics, and we will unite with all the communities that have been contaminated and that resist. Maize [is part of] our autonomy and we are not going to allow any government or corporation to contaminate it. (AJAGI et al. 2005; my translation)

Today, environmentalists and farmer organizations alike refer to transgenic contamination as a threat to indigenous communities and the culture of maize production more broadly. Fitting (2011, 114) cautions that maize activists sometimes slip into "peasant essentialism and a bounded, reified conception of culture." This is occasionally evident in statements from environmental NGOs that seem to romanticize indigenous beliefs about maize and campesino agricultural practices. On the whole, though, collaboration between environmentalists and rural communities—informed by the work of critical scientists—has produced a thoughtful and multidimensional analysis of GE maize. As the debate unfolded about GE maize after the publication of the Quist and Chapela study, some activists began to talk about global inequality along with the rights of peasant farmers and indigenous peoples. Many argued that maize contamination was a symptom of a larger problem—namely, Mexico's neoliberal agricultural policies, which have resulted in increasing food dependency on the United States (Henriques and Patel 2003). Over time, maize activists came to share the idea that transgene flow is both a symptom and symbol of Mexico's changing corn economy. Contamination is thus understood as a part of an entire complex of industrial agriculture, free trade, and oppression of indigenous people.

Those concerned about protecting maize traditions have initiated a variety of projects. Maize activists call on the state to change the systems of trade and food distribution that led to the contamination of maize, and also call on farming communities to preserve traditional seeds and agricultural practices. An agronomist who wanted to create a market for locally produced maize and traditional recipes opened a restaurant called Itanoní in the city of Oaxaca (Baker 2008). There have been numerous rural "maize celebrations," organized by small NGOs and promoters of sustainable agriculture. Mexico's Consejo Nacional para la Cultura y las Artes (CNCA, or National Council for Culture and the Arts) applied to the UN Educational, Scientific, and Cultural Organization (UNESCO) to have Mexican cuisine recognized as "patrimony of humankind" (AFP y Notimex 2005; Barcenas 2005). In a news report on the effort, Jaime Nulart of the CNCA expressed his hopes that such a designation would provide a way to confront the "grave risks" to Mexico's traditional food system, referring to GE maize as one of those threats (AFP y Notimex 2005). In 2010, UNESCO added traditional Mexican cuisine to the Representative List of the Intangible Cultural Heritage of Humanity, noting, in its decision, that "traditional Mexican cuisine is central to the cultural identity of the communities that practise and transmit it from generation to generation."[10] As each of these examples suggest, the defense of maize is a social movement of diverse organizations, taking place within multiple institutions and reconstructing the dominant cultural meaning of maize.

Forming an Epistemic Boomerang

Separate agencies of the Mexican government responded differently to public concerns about maize contamination. While environmental scientists in the government set out to gather more data about transgene flow, the Ministry of Agriculture strongly denied that there was any reason for concern. Maize activists demanded that the government halt the importation of transgenic maize to Mexico—to no effect. Finding the government largely unresponsive to their worries, Mexican maize activists, supported by international allies, decided to use the institutional mechanisms generated by the North American Agreement on Environmental Cooperation

(NAAEC), the environmental side agreement to NAFTA. Under the NAAEC, a trinational body, the CEC, works to "address regional environmental concerns," "help prevent potential trade and environmental conflicts," and "promote the effective enforcement of environmental law" (Commission for Environmental Cooperation 2011a). In 2002, petitioners asked the CEC to study the topic of GE maize in Mexico and generate a set of recommendations.

Under persistent pressure, the CEC eventually responded favorably to the petition, and created an interdisciplinary advisory group made up of scientists and other experts with a wide range of views on the risks and benefits of GE crops. The central task of the advisory group was "to examine, from different perspectives, issues related to gene flow from transgenic varieties of maize to Mexican land races and their wild relatives, and the conservation of biodiversity in this centre of origin," with the aim of producing policy recommendations (Commission for Environmental Cooperation 2011b). Eighteen experts wrote ten background chapters addressing a wide range of potential impacts of transgene flow in Mexican maize. Sixteen additional experts served on the advisory group, which comprised a diversity of expertise, including molecular biology, biotechnology, ecology, population biology, agronomy, health and nutrition, economics, philosophy, law, and political science. Most of the members of the advisory group held posts in academic or nonprofit institutions, while three represented the biotechnology industry.

When the CEC experts finally met in Oaxaca in March 2004, maize activists were beginning to clearly frame GE maize as an assault on indigenous culture and communities. In many respects, the CEC appeared to be receptive to not only environmental critiques but also social critiques of GE maize. The CEC is unusual in that it provides various formal opportunities for public participation. Still, like most environmental agencies, the CEC's approach to carrying out environmental assessments is fairly scientized, relying on experts to assess the issues. As one observer puts it, the commission endeavors to "assure an independent, neutral or scientific judgment, at the same time, taking into account the possible number of legitimate political interests" (Antal 2006). A citizen submission procedure permits citizens of the three NAFTA countries to file environmental complaints to the CEC. The NAAEC also instituted the Joint Public Advisory Committee (JPAC), made up of citizens primarily drawn from NGOs in the three member countries. The fifteen JPAC members produce consensus statements on matters relevant to CEC's activities, thus providing an additional source of civil society input (Wirth 2003).

Furthermore, the CEC typically holds a public symposium as part of the process of preparing a report and recommendations. In the case of the Mexican maize controversy, the CEC held the symposium in Oaxaca, which offered an unprecedented occasion for local critics of GE maize to bring their concerns to an international audience.

The CEC typically includes socioeconomic impacts in its assessments of environmental issues, and in the case of GE maize, the commission determined from the investigation's outset that its terms of reference would be extremely broad. The advisory group set out to consider social values and cultural identity alongside ecological, agronomic, and health issues. The inclusion of social and cultural impacts as evaluation areas drew criticism from US authorities. The US Environmental Protection Agency (EPA) formally expressed its objections to the CEC's approach to assessing the impacts of GE maize. The EPA suggested that the commission should *not* consider the effects of GE maize on indigenous peoples, and asked for the deletion of phrases such as "issues of justice and fairness in the distribution of risks and benefits among affected parties" (Redlin 2003). The CEC rejected the EPA's recommendations, keeping social, cultural, and economic impacts within the terms of reference for the report. An anthropologist and a sociologist were also selected to prepare a background paper, titled "Social and Cultural Effects Associated with Transgenic Maize Production."

Despite the CEC's clear commitment to social impact assessment and issues of justice, the application of social science perspectives to the GE maize concern did not necessarily mean providing support for the maize activists' social critique of the technology. The lengthy chapter on the social and cultural effects maintained a tone of neutrality and distance from the political conflict. The authors described the diversity and complexity of maize production systems in Mexico along with the impossibility of making a blanket assessment of the social impacts of transgenic maize (Brush and Chauvet 2004). At certain places in the report, the authors drew conclusions that contradicted the claims of antibiotech activists on some key issues, such as intellectual property and farmers' control over the seed supply. In the final pages, the report seemed to downplay the negative impacts of changes that might occur because of introducing transgenic maize, but suggested more research and public input:

Maize agriculture and Mexican society and culture are dynamic and will experience change whether transgenic maize is introduced or not, and Mexican farmers have shown that they are capable of managing their maize populations to limit or encourage change from new varieties. Whether transgenic maize will accelerate change or provoke unique or undesirable consequences in the country's maize is impossible to predict, but the possibility of this warrants further research. (ibid., 47)

The CEC scheduled a symposium to be held in Oaxaca in March 2004, in which the background papers would be presented to the public and the advisory group members would be able to privately formulate their policy recommendations. The CEC's institutionalized openness to civil society participation provided activists with a sense that it was their right to weigh in on the issue and change the direction of the CEC's plans, if necessary. The belief that the forum was intended to be open was conveyed most clearly to me in 2007, when I shared the findings of this study with one of the primary organizers of the protest events surrounding the CEC process. She expressed surprise when I referred to the CEC symposium as a "scientific meeting." Her perception (which differed greatly from that of the scientists involved, as discussed below) was that since her organization and other activists had taken the lead in bringing about the CEC assessment process, the symposium was rightfully a space for protest and citizen testimony.

Activists told me that the convergence in Oaxaca represented a turning point, when the opposition to GE maize, begun as a campaign by environmentalists, took shape as a grassroots movement driven by indigenous communities and small-scale maize producers, whose objections to GE maize were rooted in a broader critique of industrial agriculture, neoliberalism, and the loss of cultural traditions. By the end of 2003, Greenpeace and a number of Oaxacan organizations were already planning to turn the CEC meeting into a major opportunity for the movement to take a stand against GE maize. A national activist forum in defense of maize was held in Mexico City in December 2003, and there, environmentalists called on participants to converge on Oaxaca during the CEC symposium in March for a major demonstration. The day before the CEC symposium, the protesters held an alternative forum called Defend Our Maize, Protect Life.

Over 380 people are recorded as attending the CEC symposium, although Kathleen McAfee (2008), a cultural geographer who attended, observed that 380 is simply the number when the CEC lost count. Of those registered, 280 were Mexicans, and the total attendance was far larger than the CEC had anticipated (Secretariat of the Commission for Environmental Cooperation 2004). I interviewed several of the scientists who wrote background papers or served on the advisory group, and each indicated that they were unprepared for the mass mobilization that occurred in Oaxaca during their meeting. McAfee (2008, 153) noted, "It was apparent from conversations in the corridors that some council members and advisers had not previously grasped the economic importance of locally adapted maize varieties nor the centrality of maize culture to the identities of rural Mexicans. Even some of the most Left leaning of the experts felt uncomfortable with the deviation from the planned presentations and were astonished by the statements that the protesters made." One US scientist (telephone interview, May 30, 2007), a self-described "liberal academic," said that in comparison to the protesters, he felt like he was "an imperialist scientist," a fascist "in jackboots."

The mass turnout of demonstrators, particularly the unexpected testimony of indigenous people, challenged the advisory group's expectations for what the symposium was supposed to accomplish. Conrad Brunk (2006, 181), a philosopher who served as an expert on the CEC advisory group, described his recognition of a "two-way knowledge deficit" between the experts on the panel and the people who came to give testimony.

[W]hat was most evident in the campesino presentations was not only their distrust of the science, but the fact that they considered the science irrelevant to the issues of greatest importance to them. Many of the speakers explicitly voiced their distress over the fact that the background papers and the statements of the authors and panel members simply did not address their concerns. . . . The presence of the Bt transgenes in their maize was not a 'risk' posed to their cultural values, but itself already the *actualized harm*. It was already a contamination of their cultural heritage by an alien substance, present only through a strange and culturally inappropriate instrumentality. (ibid., 182)

Remarkably, as a result of this confrontation, some came to the realization that the meeting could and should serve a different purpose. Another US scientist (telephone interview, September 8, 2008; emphasis added) explained this realization to me, recalling:

I think it was Doreen Stabinsky [of Greenpeace] who said something like, "You know, this is probably making all of you scientists really uncomfortable, but this is democracy." Or something like that. She basically said, You know, this is part of life, you have to hear this, and you know, that's how I felt, I felt like *OK, we'll just change the meeting over and have it serve a different function and we'll just listen.*

By externalizing the conflict over transgenic maize to the CEC, Mexican maize producers were brought into face-to-face contact with experts who were poised to advise policymakers, and thus claimed a place in the scientific and policy debate about transgene flow. The experts involved in writing the CEC report on GE maize did not become part of the transnational advocacy network due to this interaction, but neither did they remain aloof from the diverse concerns articulated by rural people and local NGOs. Indeed, the advisory group not only listened to the activists' claims but also changed the character of its own assertions. At the symposium, the problem of GE maize was reframed for the CEC experts. What was initially treated as solely an issue of cause and effect (that is, an attempt to reach consensus on the impacts of introducing GE maize to local ecosystems and agricultural systems) became a matter of values and principles.

One scientist, the man who said he was made to feel like an imperialist scientist, said explicitly that at the symposium, he "got it"—he realized that the matter of GE maize was "not a science issue but a social issue." In his understanding, the opposition to GE maize was based on the feeling of not having a choice in the matter but instead having it imposed on rural Mexico without consultation. Although he maintained that the protests did not affect his scientific assessment of the biological aspects of GE maize in Mexico, he was a strong advocate of including the "social issues" in the official findings. Another participant told me privately that he was moved by the testimony and actions of the indigenous people who came to the CEC symposium. He thought that it had touched some other members of the advisory group as well. These comments indicate that protesters disrupted the expectations of the CEC symposium participants and led at least some of the experts to shift their framing of the issue to align more closely with activists' perspectives.

The indigenous activists' claims struck a chord with JPAC. After the meeting in Oaxaca, JPAC released a letter to the ministers of agriculture of the three NAFTA countries, stating, "What we learned from our participation [in the symposium in Oaxaca] is that the conservation of biodiversity cannot be separated from the protection of cultural diversity. A better understanding and respect for the human and social context is called for in this debate." The committee went on to observe that "the emphasis on 'scientific method' and 'science based' conclusions can work to exclude indigenous peoples." The letter also described the difficulties that the authors of the background chapters for the CEC had in "respond[ing] to the many indigenous presenters who attempted to discuss and articulate their relationship with maize as sacred, the center of life, their brother and part of their dignity and identity." JPAC concluded that the CEC lacked a balance in the advisory group's composition, and that as a result, it may promote a position that is "directly contrary to the views of the indigenous peoples in the area" (Joint Public Advisory Committee 2004).

The letter was welcomed and praised by Maria Colin, Greenpeace México's legal adviser, who was quoted in a press release as saying, "If the CEC is only guided by scientific studies, without taking into account the opinion of the citizens, a historical opportunity will be lost and this may be seen to affect its credibility" (Greenpeace México, Grupo de Estudios Ambientales, and UNORCA 2004; my translation).

The CEC Report

After the symposium, the expert advisory group developed a report and set of recommendations, published in 2004. The report's authors made an effort to distinguish between their scientific assessment and the political or value-laden concerns associated with GE maize. In a section on the process of developing the report, the advisory group acknowledged the "important social and cultural issues . . . at play," but clarified that the authors attempted to "keep those considerations distinct from the scientific evidence about health or environmental impacts" (ibid., 11). Indeed, the report is organized in such a way that "sociocultural matters" are separated from "gene flow," "biodiversity," and "health."

Scientists I interviewed indicated that even if they took seriously the concerns of protestors, their recommendations were scientifically-based. For example, among the most controversial of the recommendations was that the Mexican government should "minimiz[e] the import of living transgenic maize grain from countries that grow transgenic maize commercially" (Secretariat of the Commission for Environmental Cooperation 2004, 27). In other words, the present system of grain imports from the United States—the main source of GE maize in Mexico—must be altered. From the perspective of the local activists, the recommendation was a victory, bolstering their opposition to GE maize imports. According to one of the authors I interviewed, this recommendation was not influenced by the activists but rather stemmed from long-standing scientific concerns that maize modified to produce pharmaceutical or other inedible materials might accidentally enter the Mexican seed supply.

Nevertheless, the perspectives of local activists feature prominently in various places throughout the report. The opening paragraph of the "Key Findings" section, for example, refers to the "recent cultural memory and political history among the indigenous peoples of perceived inequity and injustice at the hands of Mexicans of Spanish origin, Americans, and powerful elites" (ibid., 14). And in the set of recommendations related to biodiversity, the advisory group urges that attention be paid to the "role and needs of campesinos, which have largely been neglected" (ibid., 28). That set of recommendations goes on to suggest that future decisions regarding agricultural technology be made with the participation of maize producers and rural communities: "Farmers of all sorts should be involved in the development of new agricultural practices from the start of the process" (ibid., 28). This proposal not only explicitly challenges the pattern of biotechnology research and development that has proceeded in Mexico thus far but also appears to blur the boundary between the expert community and ordinary citizens, as farmers are included as relevant contributors to the research process.

The sections dedicated to "Sociocultural Matters" respectfully convey the perspectives of the local activists who attended the symposium. Here, the authors appear to challenge the notion, offered earlier in the report, that scientific and cultural matters can be separated. They state: "Maize has significant cultural, symbolic, and spiritual values for most Mexicans. . . . The risk assessment of transgenic maize in Mexico is inextricably linked to these values" (ibid., 23). The section goes on to describe some of the views expressed in writing and presentations during the process of developing the report. For instance, "Many campesinos and community organizers . . . perceive GM [GE] maize as a direct threat to political autonomy, cultural identity, personal safety and biodiversity" (ibid., 23). The authors treat these perceptions as distinct from technical assessments, but equally relevant to constructing policy. They explicitly draw a parallel between the "low level of information about the fundamentals of plant genetics . . . in rural communities" and the "low level of information about rural social and cultural concerns within scientific and policy communities," both of which, they say, "frustrate the generation of scientifically sound and socially-acceptable policies" (ibid., 24).

Beyond making causal claims about the impact of transgenic maize on biodiversity, this group of experts conveyed a set of values that were shared, at least in part, by maize activists. But why? Other studies, such as the analysis done by Wendy Espeland (1998) of a water conflict involving the Yavapai Indians in the US Southwest, show the difficulty of reconciling technocratic modes of analysis with the perspectives of indigenous peoples. In this case, two things were important for reaching this rather surprising outcome. First, the typical routines for drawing boundaries between science and values were disrupted. This was facilitated by features of the organization, such as the openness to public participation and an absence of formalized rules proscribing scientists' political activities. Given such an organizational context, the disruption caused by social protest persuaded experts to adopt an alternative framing of the problem at hand and incorporate principled claims into their assessment. Second, it is likely that the scientists were able to adopt the activists' frames because the stakes were fairly low. Unlike the case that Espeland studied, in the CEC situation, the experts' recommendations were nonbinding, which may explain why consensus was achieved despite the presence of participants who were highly unsympathetic toward the antibiotech movement. Indeed, I was told by at least one informant that the probiotech members of the advisory group made little effort to involve themselves in the process—which probably would not have occurred if they expected the report to have significant consequences.

Outcomes of the CEC Process

Despite the CEC advisory group's recommendations, the Mexican government has not taken any actions to halt the importation of GE maize from the United States. To the contrary, Mexican regulatory agencies are quickly moving toward approving the commercial cultivation of GE maize. Although it is difficult to single out the most important reason for the Mexican government's lack of attention to the CEC report, certainly the nonbinding nature of the recommendations was crucial. But the report was also cast by opponents as lacking scientific authority and being slanted in favor of local activists—making it more easily dismissed.

The US and Canadian environmental agencies complained that the CEC's recommendations did not follow from the scientific information presented in the report. For example, the comments from the United States suggested that the scientific findings provided no justification for treating GE maize differently from other modern crop varieties. The United States took issue with a perceived slant in favor of the protesters, maintaining that "many of their recommendations attempt to respond solely to social-cultural perceptions of one specific group of stakeholders, while ignoring the needs of others" (Secretariat of the CEC 2004, 48). The Canadian response was also critical of the apparent influence of local activists. Environment Canada (2004, 40), for instance, observed that the recommendations were informed by "comments received throughout the process and based on the personal judgment of members of the Advisory Group"—an accusation of bias toward the protesters. Environment Canada (2004, 45) also opposed the inclusion of social considerations in the report, saying, "Risk assessment should be based solely on science. If a risk is identified, socio-economic factors may be considered when implementing an appropriate risk management strategy." Finally, while the official Mexican response did not attack the scientific basis of the report, it took issue with parts of the report that appeared to make "value judgments," particularly the line, quoted earlier, about "inequity and injustice at the hands of Mexicans of Spanish origin, Americans, and powerful elites" (CIBIOGEM [Comisión Intersecretarial de Bioseguridad de los Organismos Genéticamente Modificados] 2004, 46).

After the CEC published the report, the Mexican government largely ignored its recommendations regarding imported maize and incrementally moved toward approving cultivation of GE maize in Mexico. In many ways, the outcome of the CEC process is a disappointment to maize activists because it did not lead the Mexican government to change its approach to GE maize. One of the important outcomes of the process, however, also noted by Fitting (2011), is the greater mobilization of scientists in opposition to GE maize. Through the CEC process, activists made alliances with scientists who were already critical of transgenic crops and created a public space where those opinions could be aired. Afterward, critical scientists continued to voice their concerns about GE maize in an increasingly organized way. I was told by one of the main organizers of cultural events and protests surrounding the CEC meeting in Oaxaca that the "intention of our forum was not to influence the scientists. It was to take advantage of the scientists' forum" (maize activist, Oaxaca, March 3, 2006). He spoke highly of those activists who developed relationships with the scientific community, and seemed to conceive of their role as drawing public attention to the scientists who already were critics of GE corn:

[The groups such as Greenpeace and GEA] that fought for the CEC study . . . maintained a relationship with some of the scientists who were within that committee. The scientists . . . not only fought within the committee [for] a more honest and more realistic report, but also [shared] information. . . . Let's say, it was a feeling of having allies. . . . The government and Monsanto and AgroBIO, they were creating propaganda [in favor of biotechnology]. So we were able to confront this and say: No, look, there are other scientists who are saying the opposite, they are denouncing all that. So without a doubt, that was a contribution that [the process] represented. It was an effective alliance for our campaign and the effect it had on many people.

By the end of 2004, a small group of scientists had decided to form an organization called Unión de Científicos Comprometidos con la Sociedad (UCCS, or Union of Socially Committed Scientists) to advance the social responsibility of science. The organization was formally instituted in November 2006. Among one of the founding members, and later the president of the organization, was Elena Alvarez-Buylla, a scientist at the Autonomous National University of Mexico (UNAM), who prepared one of the early discussion papers for the CEC and has been deeply involved in research documenting transgene flow in Mexican maize. Other members of the UCCS include Antonio Turrent Fernández and José Antonio Serratos Hernández, both authors of background chapters for the CEC report. The UCCS works on a variety of issues, including climate change, forests, and energy, but agricultural biotechnology has been one of the key issues for the group from its beginnings. On June 12, 2006, Alvarez-Buylla, Turrent Fernández, and Serratos spoke at a workshop at the Mexican Congress on the proposed "special regime" to protect maize. The workshop gathered scientists, maize specialists, corn producers, and NGOs, and resulted in the "Manifesto for the Protection of Mexican Maize," signed by both scientists and activists. This represented the first joint public statement from NGOs and scientists in opposition to GE maize.[11] The manifesto calls for the government to apply the precautionary principle, support increased monitoring of GE contamination, and incorporate the perspectives of rural and indigenous communities.

Conclusion

Forming an epistemic boomerang can be a desirable option for activists when an issue has become scientized or when political authorities favor technocratic decision making. In this case, activists using the CEC process responded to the scientization of biotechnology politics in two ways: they challenged the introduction of transgenic maize in Mexico using scientific knowledge and expertise, while they insisted that other, "nonscientific" knowledge and values were equally important to the decision-making process. Activists took advantage of existing schisms in the scientific community regarding the ecological safety and desirability of transgenic crops, and did what they could to ensure that scientific critiques of GE made their way into the final report and hands of the public. An outcome of the process was the emergence of a mobilized scientific community in alliance with the maize movement.

It may seem ironic that an activist network opposed to one of the major recent developments in science and technology would choose to use scientific expertise as a resource in their struggle. It is this apparent contradiction that Ulrich Beck (1992) writes of when he describes the "risk society": we increasingly rely on science and technology to assess the risks of science and technology. Yet Mexican maize activism appears to break out of that circular process. The most striking feature of this case is that maize activists actively challenged the dominant framing of GE maize as a risk-management problem, instead showing it to be a far-reaching social, cultural, and economic dilemma that experts only partially comprehend. And indeed, their success in transforming the CEC experts' mode of assessment of the issue may ultimately be one of the most significant outcomes of this struggle.

When the CEC process was initiated, the emphasis was on the scientific analysis of the effects of transgenic maize on biodiversity. Over the course of the CEC process, though, activists pushed for a broader conception of what constituted relevant knowledge for the evaluation of the impacts of transgene flow in Mexico. The experts that maize activists confronted in public collectively shifted the character of their claims about maize. In their final report and recommendations, they not only made causal claims, as expected in an environmental assessment, but also adopted elements of the activists' framing of the issues, referring to local cultural values and the need for participatory decision making.

On the one hand, the experts' principled support of the activists' positions can be seen as a victory for maize activists; on the other hand, the inclusion of such allegedly unscientific statements in the CEC report provided an easy justification for government authorities and other critics to reject the experts' conclusions. The CEC report was met with overwhelming pushback from Mexico's trading partners, who condemned the expert recommendations for reflecting something other than a narrowly technical assessment. As in the case of Roundup Ready alfalfa in the United States, a window of opportunity was opened to officially consider the protection of alternative systems of agriculture, but was quickly closed.

Despite the success of the advocacy network in bringing attention to the socioeconomic and cultural implications of transgenic maize, the political outcomes have not been in the activists' favor. The Mexican government has experienced no compelling pressure to implement the commission's recommendations, and experimental field trials have proceeded despite objections from the UCCS and other scientific experts. This suggests a need to systematically consider the circumstances under which epistemic boomerangs may be more likely to produce tangible political changes. Even where activists do gain expert allies, this will fail to affect government policy if states cast the resulting scientific advice as lacking credibility. This is likely in contexts where credibility is assumed to require distance from contentious politics. States and industries resisting the pressures of epistemic communities benefit from perpetuating the notion that scientific advice must not be "tainted" by social concerns that are of central interests to activists. Presently, then, activists dealing with the complex problems resulting from developments in science and technology face a dilemma. They depend on scientists, to both produce the answers they need and lend credibility to their concerns. Yet when scientists appear to share the values and political ideas of activists, their work may be more easily discredited by opponents. Does this mean that activists wishing to back up their positions with scientific evidence ought to discourage scientists from making principled claims on their behalf? Such a route might have the short-term benefit of protecting the perceived credibility of the scientific data. In the long run, however, such distinctions perpetuate a hierarchy of science over values that ultimately exclude ordinary citizens from decision making about technological change.

The next chapter turns to a different way that maize activists have participated in scientific debates about transgenic crops. NGOs and rural organizations have collaborated to monitor the presence of transgenes in maize harvests across the country. As in the case of the CEC's report, these efforts to generate new scientific assessments of genes out of place have been controversial, facing criticism not only from scientific experts but also within the network of maize activists. In both of these examples, scientific debates are not merely the backdrop to the social conflicts, but central to the formation of political solidarities and the articulation of grievances about threats to rural livelihoods. These struggles challenge simple distinctions between science and politics, or global and local discourses. As in the case of the CEC's report, the next chapter illustrates how notions of rational risk assessment are being reconstructed by challengers to better represent the full complexity of grievances about the threats to maize.

4

The Politics of Biosafety Monitoring

Since the early 1990s, biodiversity protection has been a key international priority, enshrined in UN agreements such as the CBD and its supplement, the Cartagena Biosafety Protocol.[1] Both international agreements establish the importance of *monitoring* biodiversity in order to protect it. These agreements provide little guidance on how to monitor or how such monitoring might lead to biodiversity protection. Nevertheless, the monitoring of GE plants released into the environment is now widely considered to be an important component of environmental governance in areas with high levels of biodiversity (Pearson 2009).

Tracking the spread of GE material through a population is no easy task, though, particularly since plants containing transgenes look no different from other plants and must be analyzed at a molecular level. Allison Snow (2009, 569), a plant evolutionary ecologist who is prominent in debates about transgene flow, describes monitoring for transgenes in Mexico as a "daunting task that requires teams of experts from several disciplines." Since the 2001 surprise discovery of transgenic material in maize growing in Oaxaca (Quist and Chapela 2001), scientists have developed sophisticated methods for detecting contamination and monitoring transgene flow, confirming the presence of transgenic sequences in Mexican maize landraces (Dyer et al. 2009; Mercer and Wainwright 2008; Mezzalama, Crouch, and Ortiz 2010; Piñeyro-Nelson et al. 2009a, 2009b; Serratos-Hernández et al. 2007).[2]

Both scientists and maize activists have criticized the Mexican government's monitoring capacity. In 2009, Mexico's UCCS sent a letter to President Felipe Calderón, urging him to ban field releases of commercial GE maize varieties, and "increase to a level of scientifically-sound efficacy the infrastructure necessary to monitor and independently evaluate seed and grain entering Mexico." In an addendum, synthesizing the major uncertainties and risks associated with GE maize, the letter repeatedly drew attention to the current limitations on biomonitoring. For example, the authors pointed out that the detection of transgenes "has been hindered by the lack of public access to reliable sequence data on all the recombinant constructs that could be involved," and argued that "the certified standard methods to detect transgenes in hybrid maize in the USA and Europe are inadequate for monitoring transgenes in landrace varieties." Furthermore, the letter asserted, the Mexican government has not shown itself up to the task of monitoring GE maize. "The biosafety and biomonitoring infrastructure provided by the Mexican authorities is insufficient: there is only one national laboratory certified for GM maize detection. . . . [Moreover,] since 2007, a national network of GMO monitoring laboratories has been created, but it is still not operating and it is not clear how it will be supervised by the Government" (Unión de Científicos Comprometidos con la Sociedad 2009). More generally, some say that even when scientific monitoring takes place, "the politically sensitive nature of this information has made it difficult for researchers to publish their findings" (Snow 2009, 569).

Participants in the Network in Defense of Maize, an eclectic group of research and advocacy NGOs, grassroots development organizations, indigenous community organizations, and campesino groups, have taken a more radical position on the Mexican government's efforts to monitor GE maize. In a 2009 declaration, the network "repudiate[d] government monitoring of peasant cornfields, because it is a pretext for eliminating even more peasant seeds" (Red en Defensa del Maiz 2009). In an editorial in the *La Jornada* newspaper, Silvia Ribeiro, an ETC Group staff member, further explained the activist network's position against government monitoring. Ribeiro, a key participant in the Network in Defense of Maize, criticized the government's network of monitoring laboratories on a number of grounds. The editorial expressed distrust in the laboratories, suggesting that they might hide or manipulate findings that do not support the government's favorable view of biotechnology. Ribeiro also articulated suspicions that discoveries of contamination by the monitoring network would give the government an excuse to force campesinos to give up their maize landraces and switch to commercial seeds. Furthermore, she contended that monitoring could be used against farmers, as a means of discovering who was using patented GE seeds without paying for them—a reference to Monsanto Company's legal actions against US and Canadian farmers. Ribeiro (2009) concluded that the government's efforts to monitor GE maize, while giving the appearance of concern for biosafety, would in fact be used against campesino agriculture, to the advantage of multinational seed companies.

As these critiques by the UCCS and Network in Defense of Maize indicate, the monitoring of genes out of place is not a politically neutral endeavor, despite its basis in scientific methods. Who monitors, where they monitor, the methods of analysis used, the publication of findings, and the ends to which the results are used are all points of contention. Concerned scientists have criticized the Mexican government's actions and carried out research on transgene flow, allying themselves with social movement organizations that are opposed to GE maize. At the same time, some activists, primarily Ribiero and others in the Network in Defense of Maize, have questioned the very premise and purpose of monitoring. Distrustful of research by government scientists, activists have taken on the task of monitoring GE maize themselves.

In this chapter, I show that debates about monitoring reveal not only a conflict between challengers and the Mexican government but also two distinct political orientations among maize activists in Mexico. It is not uncommon for social movements to have various factions that differ in their relationships to dominant institutions and favor different strategies for pursuing social change. Here, I explore how such factions take shape in relation to scientific knowledge work—in this case, environmental monitoring. I begin with a discussion about the growing role of diverse publics in defining risks and participating in environmental monitoring efforts. I then turn to the Network in Defense of Maize's research effort in three parts: first, the role of the monitoring project in mobilizing rural communities in defense of maize; second, the network's confrontations with institutional science; and finally, the implications of the network's most controversial claim about the impacts of genes out of place. Throughout this analysis as well as in the conclusion, I focus on the ways that activist-led monitoring efforts contribute to the formation of political solidarities and, in turn, reflect commitments to particular strategies for social change.

Reconstructing Risk

The creation of monitoring programs is a typical way that governments deal with public concern about the potential consequences of technological change. Though systematic monitoring of quantifiable indicators, questions about the desirability of technological developments are scientized. Monitoring is everywhere, from radiation sensors, to systems for reporting the adverse effects of pharmaceutical drugs, to air and water pollution detectors. The ubiquity of monitoring systems is symptomatic of what Beck (1992, 19) calls, as mentioned earlier, the risk society—a society that faces the "risks and hazards systematically produced as part of modernization" by asking how "they [can] be limited and distributed away so that they neither hamper the modernization process nor exceed the limits of that which is 'tolerable.'"

The appeal of monitoring projects derives, in part, from the invisibility of many environmental and health hazards, such as radiation and toxic pollution. The risks associated with contemporary science and technology

induce systematic and often *irreversible* harm, generally remain *invisible*, are based on *causal interpretations*, and thus initially only exist in terms of the (scientific or anti-scientific) *knowledge* about them. They can thus be changed, magnified, dramatized or minimized within knowledge, and to that extent they are particularly *open to social definition and construction.* (ibid., 22–23; emphasis in original)

Monitoring, in this context, is one of the crucial tools for establishing that invisible risks actually exist. As scientific knowledge becomes ever more essential to demonstrating the existence of invisible risks, nonscientists have taken an increasing role in producing knowledge about environmental and health threats, using tools like air-monitoring "buckets" (O'Rourke and Macey 2003; Ottinger 2010; Overdevest and Mayer 2008) and techniques of "popular epidemiology" (Brown 1987, 1992; Clapp 2002).[3] Participation in monitoring activities is one of the ways that ordinary citizens engage in scientized political debates and challenge the claims made by experts about technological risks.

The risk society thesis associates the public scrutiny of science with the recent period of late industrial modernity.[4] Yet research in the Global South shows that "public dissent and lack of trust in expert institutions is not so new, and not uniquely a feature of late industrial modernity in the West" (Leach and Scoones 2005, 20). From critiques of colonial science and technology to popular resistance to development plans such as big dams and conservation projects, communities in the Global South have long been attuned to the problems generated by scientific modernization. Development scholars Melissa Leach and Ian Scoones (2005) indicate that globalization processes have diminished the significance of North–South distinctions with respect to risk consciousness (see also Fairhead and Leach 2003). Leach and Scoones (2005, 31) maintain that "in the context of contemporary globalization it is not appropriate to characterize 'late industrial society' as specific to certain geographical locales," because risk definitions are "quintessentially locked into globalized scientific and policy fields" that transcend North–South divides. They (ibid., 36) conclude that "experiences of extreme vulnerability and marginalization from science/policy processes are common to groups of people in Europe and the USA as much as in Asia and Africa, while the latter, too, have their groups of 'scientific citizens' contesting official perspectives in Euro-American, reflective, 'risk society' style."

Consistent with these observations, anthropologist David Hess (2007) attributes the growing role of diverse publics in monitoring risks and critiquing scientific and technological trajectories to processes of globalization. The current period of globalization is characterized by the growth of supranational and transnational governance bodies, increasing economic integration, and changing patterns of migration and cultural exchange. In this period, there are, on the one hand, growing pressures for researchers to align their work with industrial priorities. On the other hand, a countervailing process has developed, in which there is increasing scrutiny of science "from below" (see also Moore et al. 2010). Hess (2007, 47) uses the term *epistemic modernization* to refer to

the process by which the agendas, concepts, and methods of scientific research are opened up to the scrutiny, influence, and participation of users, patients, non-governmental organizations, social movements, ethnic minority groups, women, and other social groups that represent perspectives on knowledge that may be different from those of economic and political elites and those of mainstream scientists.

Epistemic modernization is the outcome of several institutional changes, including: the diversification of the social composition of science, including previously underrepresented groups and international participation; the growth of community-oriented research projects, such as "science shops" and participatory public health research; the growth of social movement organizations that challenge the epistemic authority of science; and efforts by dissident scientists to ally themselves with social movements and construct alternative institutions such as "concerned scientist" NGOs. I would add to this list the creation of epistemic boomerangs—activist networks that call on scientific experts to provide recommendations and advice to governments that are otherwise unwilling to respond to local grievances, as described in the previous chapter.

As epistemic modernization occurs around the globe, it is not a homogeneous process. When the institutional changes discussed above take place, they do so in ways that reflect the specific political, economic, and cultural features of a society. In societies where indigenous peoples are struggling to resist the further loss or transformation of their cultures, challenges to the epistemic authority of science come not just from dissenting scientists, social movements, and an engaged national citizenry but also from oppressed communities that actively resist "modernization" projects that are imposed in the name of technical progress and economic development. In a similar sense, although the concept of risk travels internationally, it does not necessarily operate in the same ways across cultures, particularly as local social movements creatively appropriate the concept to meet their own needs and ideas (Leach and Scoones 2005; Fairhead and Leach 2003). Beck (1992, 72) rightly observes that risk consciousness is determined by and oriented toward science—not lay or traditional knowledge. That is, "risk" is a thoroughly modern and scientific way of perceiving the world. As diverse cultures encounter scientized global and national institutions, risk statements become the dominant mode of expressing concern about science and technology. Risk consciousness, though, may be combined with the knowledge and perspectives of nondominant groups in society, in a process that Hess (1995, 41) refers to as *reconstruction*.

In this context, reconstruction refers to the "ways in which different groups actively reconstruct sciences and technologies by positing alternatives consciously linked to their social identity" (ibid., 41). In particular, Hess points to indigenous peoples' resistance struggles contesting "cosmopolitan" technologies. Against simplistic understandings of culturally "pure" societies either resisting or being destroyed by outside influences, Hess (ibid., 220) argues that it makes more sense to "think of science and technology as helping to create new hybrid cultures." Some technologies primarily work in ways that destroy or colonize indigenous societies—such as dams, military testing sites, and modern agricultural technologies. Other technologies can become tools of resistance, such as media technologies, medical technologies, and useful infrastructure— not to mention environmental-monitoring technologies. Furthermore, resistance can be proactive, creating new knowledge, practices, and technologies that enable communities to resist agents of destruction. As an example, Hess (ibid., 233) highlights "grassroots development" projects, which typically aim to generate bottom-up, alternative approaches to social and technological change that are controlled locally. Agroecology projects supported by civil society organizations and other efforts to revitalize traditional agricultural knowledge in Mexico are examples of grassroots development.

Processes of reconstruction are evident in the work of NGOs and rural community groups connected through the Network in Defense of Maize.

Participants in this network, spread out across Mexico, are often involved in other community-based struggles related to agriculture and economic survival. This network figuratively calls itself a *petate* (pronounced pe-TA-tay), the name for a woven mat made out of palm leaves. It symbolizes the dense network of people forming the base, or grassroots, of the movement in defense of native maize. Borrowing the maize activists' self-description, I refer to the Network in Defense of Maize as "the petate" here. The main settings for direct interaction between the multiple groups participating in the petate are national and regional "forums in defense of maize," coordinated and facilitated by a variety of different NGOs. Since 2002, these gatherings have been sites for information exchange and the development of a collective identity with respect to native maize contamination. The forums have also inspired scientific inquiry, in the form of research efforts led by activists. The petate worked with farmers and used commercially available methods of genetic testing to carry out a geographically extensive study of maize contamination. Environmental organizations and activists have used genetic tests to identify transgenic material in foods and plants around the world. Nevertheless, in adapting those monitoring methods to Mexican maize systems, indigenous cultures, and sustainable rural development projects, the effort took on new meanings and purposes.

Members of the petate are clearly attuned to scientific debates and frequently make risk statements. Activists from organizations ranging from transnational NGOs to small groups representing indigenous communities make claims about the hazards that GE maize is likely to bring. The petate's risk claims are shaped by and oriented to scientific knowledge, yet they also frequently draw on the knowledge of maize producers and incorporate ideas about maize that stem from indigenous cultural beliefs, such as the idea that maize has a soul. Just as significantly, claims about risks also consciously reflect the petate's rejection of institutional politics and political struggle at the national level in favor of grassroots cultural transformation. In a context of great uncertainty about the impacts of trangene flow in native maize, these political and cultural commitments led petate members to view some risk claims as more plausible than others. Those not in the petate, though, regard some of the claims that the network makes as unscientific and even irresponsible. My aim here is not to assess whether the petate's claims are factually correct but rather understand their complex institutional and organizational context and the effects they have on the mobilization of alternative pathways in Mexican agriculture.

The First Petate Study

The idea for an activist-led study took shape at the first Meeting in Defense of Maize in Mexico City in January 2002. One of the organizers
of that meeting told me that she and other representatives of NGOs in
Mexico City had met after the news of the discovery by Quist and Chapela (2001) emerged, and they decided to facilitate greater public debate
about the subject. They especially sought the participation of campesinos
and indigenous people, since they would be the most affected by changes
to maize production. Over four hundred people attended the Meeting in
Defense of Maize, where the findings of the newly published Quist and
Chapela study as well as a related investigation carried out by the Insti-
tuto Nacional de Ecología (INE, or National Institute of Ecology) were
shared and discussed.[5] Both of those studies found GE maize growing
in Mexico, but focused primarily on one region of Oaxaca and parts of
Puebla, a neighboring state. At the forum, small-scale maize producers
from many regions of the country said that they also wanted to know
whether their maize was contaminated. Therefore, in the months and
years that followed, a coalition of organizations based in Mexico City
responded to this need for further investigation by seeking funding for
independent tests for contamination.

Several organizations based in Mexico City—CECCAM, ETC Group,
CENAMI, and the Centro de Análisis Social, Información y Formación
Popular (CASIFOP, or the Center for Social Analysis, Information, and
Popular Training)—jointly coordinated the research, with CECCAM taking the lead on much of the scientific analysis (I refer to people from this
group of NGOs as the "petate study coordinators"). From the perspective of these organizations, the research effort was intended to initiate
a "process" in rural communities that would be "totally participatory"
(petate study coordinator, Mexico City, December 9, 2005). Sites for the
research effort were selected partially on the basis of their potential for
generating organization and mobilization at the grassroots level—that is,
communities where it seemed to the coordinators that the study would
lead to ongoing involvement in the defense of maize.

Because so little was known about transgenic crops in rural communities, the coordinators organized workshops with the idea of preparing
people to go back to their communities to teach others what they had
learned. But the coordinators found that it was not easy to explain genetic
engineering to the rural population. Education levels in rural Mexico are

low, so there is little to no understanding of plant genetics. As one petate study coordinator (Mexico City, December 9, 2005) described it:

At first we tried in every way to explain what transgenic crops are . . . and no one understands what "transgenic" means. It's not very easy because it is something that can't be seen, and the campesino and indigenous experience has always been with things that can be seen and touched. . . . We learned lots of ways to explain it until we found a way that more or less gave the idea that it is something that your eyes aren't able to see, and it's not because you don't have experience but rather because they put something really deep inside the heart of maize. And so this helped. . . . But what we really learned was that it wasn't so necessary to explain it. Explaining how it is done and everything was the least of it. [It was sufficient] to explain that a transgenic plant was made for the benefit of a few who made it property, when maize belongs to everyone or to no one. . . . It was sufficient [to discuss] the risks that they produced and the benefits that the companies claim. It wasn't necessary to explain what it was [at a molecular level], and though we could talk ourselves hoarse, we saw that was kind of a useless effort.

In a sense, within the activist network itself, there is an expert–layperson divide. There is a clear distinction between those who understand the workings of genes and genetic engineering and those who do not. The former, working in advocacy and sustainable rural development organizations, could be considered "lay experts" (Epstein 1996). I spoke to a variety of people who worked for small organizations that aimed to assist rural communities with improving their agricultural livelihoods, and asked how they went about explaining the issue of transgenic crops to campesinos. Many use teaching materials such as drawings and diagrams to describe the cells of a plant and how DNA is spliced. At the same time, these educators highlighted the high level of expertise that the campesinos have with respect to the maize that they cultivate, sometimes using the phrase "campesino science" (Marielle and Peralta 2007) and indicating that producers know much more about maize than they do. The sentiment that maize producers are the true "maize experts" is widespread in Mexico's maize movement, reflecting an effort to place local knowledge on an equal footing with scientific knowledge.

For the study, coordinators and rural participants took samples from two thousand plants in 138 communities across Mexico. The analyses were done using commercial kits made by Agdia Testing Services, a company based in the United States that has offices around the world, including three laboratories in Mexico. The assay employed by the test kits is used to detect proteins associated with particular commercially available transgenic crops. Some of the initial assays were done in the communities themselves with the technical assistance of unnamed biologists at UNAM

(ETC Group 2003). The NGOs that coordinated the study said that scientists aided them in certain aspects of it, but that those scientists wished to remain anonymous. As the petate study coordinator (Mexico City, December 9, 2005) quoted earlier observed:

Certain scientists have participated and have helped us, and also tipped us off to discussions that were going on in the scientific community. . . . [But] we have to protect them because they have to earn a living. There are some of those scientists who have decided to speak out and they have been run out of their positions, and suffer repression.

Later, the activists switched to sending the samples to be analyzed, using the same test kits, in a commercial laboratory in Mexico.

An analysis of the samples revealed transgenes in plants growing in thirty-three communities, spread across nine states (ETC Group 2003). The study conducted by Quist and Chapela along with those conducted by agencies of the Mexican government had only focused on a small region of Oaxaca as well as a few communities in the state of Puebla, so this was the first time that such a large geographic scope of transgene flow in Mexico was indicated. The transgenes detected were those developed and patented by Monsanto, Novartis, and other biotechnology companies. Perhaps most alarming was the presence of StarLink genes, which have never been approved for human consumption.[6] When the results of the first round of activist-initiated testing of maize samples were compiled, the organizations involved wrote a five-page press release explaining the details of their research and presented their findings at a press conference. In October 2003, the findings were reported in at least two major Mexico City newspapers (*La Jornada* and *El Universal*) and Spanish-language newswires (Reuters 2003; Efe News Services 2003; Perez 2003). The organizations directing the study also printed up its results on posters, to be distributed to rural communities. The NGOs announced the findings internationally and circulated an "open letter" to the Mexican government and international community, demanding that immediate action be taken to address the widespread contamination. The letter gained hundreds of signatures from organizations globally.

In interviews with academic and government scientists about the GE maize controversy, I asked what they thought of the NGO-led study, and the response was always that they needed to know more about whether it was methodologically sound in order to have confidence that the findings were valid. It is not unusual for researchers to use commercial laboratories for plant sample analyses; scientists used a laboratory called Genetic ID in peer-reviewed research on transgenes in Mexican maize (Ortiz-García et al. 2005). Some scientists, however, pointed out the shortcomings of the test kits and questioned the use of this particular laboratory. For example, one scientist in an environmental agency said she did not know of the laboratory and that she could not be certain that the study was done with scientific rigor (regulatory official, Mexico City, November 7, 2005). Another government researcher made a similar observation, explaining "I don't want to say that in certified labs everything goes perfectly; they also have their problems. But the probability of having an erroneous result is less [in a certified laboratory]. . . . The work that the civil society organizations are doing is valuable, but I worry that they are making decisions based on results that are not very solid" (government researcher, Mexico City, November 3, 2005).

Confronting Institutional Science

Ecologist Sol Ortiz-García, director of the biosafety program at Mexico's INE, has been leading an effort to monitor the presence of transgenes in Oaxaca and the surrounding states ever since Quist and Chapela made their startling discovery of transgenes in samples of Oaxacan maize. Initially, INE research confirmed that native maize was indeed contaminated, but the research report was never published in a peer-reviewed journal. In 2005, Ortiz-García and her colleagues published new findings in the *Proceedings of the National Academy of Sciences*, a US scientific journal. They reported that they found no evidence of transgenes in samples of maize over the course of two years (Ortiz-Garcia et al. 2005). The authors did not dispute the earlier findings by Quist and Chapela (2001) but rather argued that any previous contamination had been reduced to undetectable levels. This could have occurred, the authors indicated, because maize producers had become more careful about not planting seeds imported from the United States. The researchers thought that maize activists would welcome this report since it was essentially good news (Fitting 2011, 59–60). Yet the INE study undermined the claim frequently made by antibiotech activists that contamination is irreversible, and this was hard for some antibiotech activists to believe.

When the INE study was published, critics of Quist and Chapela's study used it to contend that previous findings of transgenes were simply the result of sloppy research, and that there had *never* been GE contamination. Despite the careful wording of the study by Ortiz-García and her colleagues, it was widely reported in the press that it proved that Quist and Chapela had been wrong all along, and that there was no basis for contamination claims. *AgBioView*, a probiotech email list, vehemently promoted this view (Morton 2005; Prakash 2005). One editorial asked, "Why are Ortiz-García et al. assuming that transgenes were present before? The authors admit that no one has really proved that transgenes are present in Mexican maize. . . . Why not admit the introgression most likely never happened in the first place?" (Morton 2005). The sentiments of the biotechnology industry are clear in another *AgBioView* editorial, gleefully titled "Duh . . . No GM Genes in Mexican Corn," written by vocal biotech proponent C. S. Prakash. Without questioning whether one can "positively confirm" the absence of something, Prakash (2005) maintained that the study "positively confirmed that biotech traits are not present in native landraces of maize in Oaxaca."

ETC Group, which frequently uses press releases and newspaper editorials to comment on the GE maize controversy, used the petate's findings to respond to the public debate about the Ortiz-García study. In a press release, the organization criticized not only "the biotech industry's self-serving interpretation of the study" but also the investigation itself, on methodological grounds (ETC Group 2005). The ETC Group critique of the INE study relied in important ways on the results of the activist-initiated study. First, the activists minimized the novelty of Ortiz-García's negative findings, arguing that they were not surprising because some of the samples were taken in communities where the activist network had already done testing and found negative results. Furthermore, the INE study did not account for the findings of the activist-led study. The press release quotes Baldemar Mendoza, representing the Unión de Organizaciones de la Sierra Juárez de Oaxaca (UNOSJO, or the Union of Organizations of the Sierra Juarez of Oaxaca), an organization that promotes sustainable rural development for the indigenous Zapotec communities:

"We took samples in 3 of the 18 communities that the new report mentions (San Juan Ev. Analco, Ixtlan and Santa Maria Jaltianguis) and our results were also negative in those three communities." . . . Mendoza also points out, "The new study doesn't refer to any other part of Mexico where contamination has been found but some in the media are already making the false claim that 'there is no contamination in the whole state of Oaxaca or even all of Southern Mexico.'" (ibid.)

The press release went on to reiterate the findings of the activist-led study, indicating that contamination had been found in nine Mexican states— areas that remained unexamined by Ortiz-García and her colleagues.

In addition to critiques based on its own research, ETC Group also pointed out what it viewed to be other methodological flaws. The negative findings, ETC Group argued, "could demonstrate that the testing technology is every bit as unreliable as the genetic transformation technology—since the behavior of transformed genes isn't always predictable." The organization suggested that the authors "erroneously inflated their sample size, thus giving their results an unwarranted appearance of accuracy." Finally, the article indicated that the commercial testing companies that conducted the analysis use conservative procedures that would not be able to detect low-level contamination, in contrast to the tests used in previous studies (ibid.). This critique, and the activist-initiated study on which it was based, were widely circulated on antibiotech and sustainable agriculture Listservs. In an August 10 statement, Quist and Chapela also indicated that they had found "troubling methodological and technical problems" in the study, and that they were writing a rebuttal (to date it has not appeared).[7]

Not all maize activists shared ETC Group's view of the INE study. One grassroots development organization sent maize samples to the INE for analysis on more than one occasion. The INE used the commercial laboratory, Genetic ID, to test the samples, because the institute had not yet developed its own certified laboratory. A representative from the grassroots organization said, "I think we can be confident [in the results of the INE's sample analysis]. . . . We are not worried that the result came through INE" (agroecology promoter, Mexico City, December 15, 2005). In particular, the similarity in the findings of the activist-led and INE studies (that is, no contamination found in the same areas) strengthened the organization's confidence that the results published by Ortiz-García and her colleagues were correct. What ETC Group viewed as a weakness of the study, in other words, could be understood as a sign of its credibility.

The difference in interpretation in this instance is consistent with broader differences among maize activists in their level of trust in the government to produce credible knowledge about GE maize. Indeed, during the time I was in Mexico to study this controversy (2005–2006), it was evident that within the broader maize movement, the petate was developing a unique stance with respect to the government. While other activist organizations might work for policy change at the national level, those organizations and communities in the petate would focus on community-level changes in agricultural practice. Many of those involved in the activist-led study were of the view that the government was unwilling to act on *any* scientific information that indicated the presence of transgenes in native maize. The experience with the CEC process, in which the Mexican government simply refused to act on the recommendations made by the expert panel, contributed to this perspective. Take, for example, this statement made by an advocate for campesino agriculture:

The majority of the scientific community has pronounced itself against the commercial and scientific culture of transgenic crops, to stop the imports and to renegotiate NAFTA in questions of basic crops. . . . We face the possibility of a biological and cultural disaster in terms of the maize [yet] the government does not act in consequence. (petate study coordinator, Mexico City, October 20, 2005)

Some research scientists I interviewed shared the opinion that the Mexican government does not listen to the advice of scientists who raise concerns about the environmental safety of GE maize, and this view has been publicly voiced by the UCCS since it was founded in 2006. Despite formal language advocating science-based policy on GE maize, there is a general tendency on the government's part to ignore or diminish the importance of scientific research that would discourage the use and development of GE crops. The agriculture ministry (SAGARPA), in particular, has prioritized trade over calls for further environmental safety assessments. While the scientific debate has continued, with new findings of contamination published in prominent journals (Dyer et al. 2009; Piñeyro-Nelson et al. 2009b), it seems to have little effect on the regulation of GE maize imports and cultivation. The Mexican government decided to proceed with the experimental cultivation of GE maize in 2009.

In interviews conducted in 2005 and early 2006, activists in the petate said that they had begun a "second phase" of the struggle. They would no longer attempt to influence legislation or regulatory decisions, since they did not foresee any success through that route. The petate does continue to periodically publish pronouncements on national issues—such as the one denouncing the Mexican government's monitoring of GE maize, described in this chapter's opening—and collect signatures from supporting organizations around the world. Still, they do little direct advocacy work on national policy issues. The participants in the petate aim to change agricultural production practices on a local, field-by-field, community-by-community scale. For example, they advocate the cultivation of native maize varieties, seed saving, and the avoidance of seeds of unknown origins. The 2009 declaration, cited in the first pages of this chapter, called on "all indigenous and peasant communities" to "defend native seeds and continue planting, storing, exchanging, and distributing their own seeds, as well as exercising their right over their territories and preventing the cultivation of transgenic maize" (Red en Defensa del Maíz 2009).

Activists in this network cite a number of factors that contributed to their decision to forego national-level engagement. These include the government's refusal to act on the San Andrés Accords for indigenous rights, the unsatisfactory outcome of the 2003 El Campo no Aguanta Más movement, the passage of the LBOGM, and finally, the failure to provoke a government response to the findings of GE maize growing across Mexico. Many rural communities involved in the petate are closely affiliated with the indigenous rights movement and the CNI, through which they have pursued political, economic, and cultural autonomy without waiting for legal reforms that have been slow to materialize. Maize activists in the petate did not want to negotiate with what they called "bad government." One Mexico City–based activist, for instance, described how she came to feel disillusioned with the legislative process after the LBOGM's passage:

> Evidently the representatives and the senators didn't think at all, because the law was passed practically by consensus . . . and there had been a lot of discussion because there was tremendous mobilization around the topic in Mexico. . . . Artists, intellectuals, scientists, all the campesino organizations, even the unions, in other words, all the sectors in Mexico protested against the [proposed] law. And anyway it changed absolutely nothing. There were some cosmetic changes. That is our opinion—there are other organizations that think that they achieved two little things in the law—[but] we think that nothing was gained. (petate study coordinator, Mexico City, October 4, 2005)

This convinced her that there were no remaining opportunities to create change through the national state.

There are many organizations and communities involved in the maize movement that have not directly participated in the petate, such as Greenpeace Mexico, GEA, and UNORCA, each described in the previous chapter, as well as numerous other environmental and campesino organizations that continue to view the national government as a target. They sign petitions, hold demonstrations, denounce the government's support for GE maize, and promote alternative visions of Mexico's agricultural future. Some advocacy organizations, particularly Greenpeace Mexico and GEA, have had access to policymakers and elite bureaucrats within the regulatory agencies, including state scientists. For example, in 2005, activists from both organizations were regular participants in a yearlong series of monthly meetings aimed at building capacity among state agencies to implement the LBOGM. Although activists from those organizations are highly critical of the new regulatory rules for GE crops and are dismayed with the government's promotion of biotechnology, they continue to view political pressure and participation as effective and necessary ways to produce social change.

As I will discuss further below, for those in the petate, the experience with the activist-led environmental monitoring precipitated a critique of scientific authority and a new articulation of what it means to have credibility. In this sense, the petate has distanced itself from the knowledge politics of environmental advocacy groups that tend to rely on science and engage in mainstream scientific debates. Not surprisingly, the organizations that have access to state elites and the institutionalized political process tend to seek policy change through scientific argumentation, in keeping with the scientization of politics. For instance, science and scientific credibility are central elements of Greenpeace activism in Mexico, combined with other approaches such as attention-getting demonstrations. A Greenpeace Mexico activist (Mexico City, October 26, 2005) acknowledged that credibility was a concern for Greenpeace, and

for that reason we are very strict in the analysis and in the search for information that we publish, we are very rigorous in this sense, that is, let's say, Greenpeace has capital. [The organization] has constructed capital in credibility, which has been constructed over 32, 33 years at the international level and it's something that we seek to maintain and grow.

He added that for Greenpeace, credibility means both making sure the information they use is accurate but also "telling the truth," unlike probiotech actors that "hide information because they have commercial, economic interests." As such, one of the strategies of Greenpeace Mexico has been to publicize scientific information on the issue of transgene flow in maize and criticize the government when its decisions appear to be biased toward corporate interests.

In contrast, activists in the petate, working at a greater distance from government institutions, express a more ambivalent view of science. Organizers of the activist-initiated study have emphasized that their research process puts campesinos, not scientists or professional activists, at the heart of the movement. Maize producers are central, not just as victims of transgenic contamination, but also as active participants in processes of knowledge production. Some prominent voices in the petate criticize the practices of institutionalized science for not meeting the needs of campesino communities. For instance, one rural Oaxacan indigenous rights activist (Oaxaca, March 6, 2006) explained:

The problem with the scientists is that they need to have proven information to fall back on. We, well, we work a different way and what interests us is to share information that enables people to make decisions quickly. . . . This information isn't useful if we don't circulate it right now.

The independent, NGO-led study is thus favored over the slow pace and seeming ineffectiveness of formal scientific research. In this case, the justification for doing their own research is that campesino communities may be harmed by long delays waiting for the peer-review process and formal publication.

Maize Malformations

The petate's targets, strategy, and orientation toward scientific authority are particularly well illustrated by its controversial claim that transgenic contamination is causing maize deformities, thereby threatening food production. In 2003, when the petate published the findings of its study, it not only reported widespread GE contamination; it also indicated that deformed plants have been found and, moreover, tested positive for transgenic DNA. Although the activists acknowledged that they did not have evidence of a correlation between transgenes and malformations, they believed that the recent appearance of high frequencies of the misshapen plants could be caused by the influx of transgenic maize. The press release stated that "deformed plants have been found in the states of Oaxaca and Chihuahua that have tested positive for the presence of GM products" (ETC Group 2003). Mendoza said, "We have seen many deformities in corn, but never like this. One deformed plant in Oaxaca that we saved tested positive for three different transgenes. The old people of the communities say they have never seen these kinds of deformities" (ibid.).

Questions about deformed maize plants began to gain momentum—as well as attract controversy—around the time of the CEC symposium in Oaxaca, held in March 2004 and discussed in the previous chapter. As many people I interviewed recalled vividly, some protesters at that event expressed concerns that contamination was physically harming maize plants. In particular, one indigenous rights activist from Oaxaca brought the deformed stalk of maize that had tested positive for three different types of transgenes (also reported in the press release of 2003). After the 2003 press release and CEC symposium in 2004, more malformations were spotted in corn plants across Mexico. Speculation about whether they were the result of genetic contamination became a theme of activist forums and farmer workshops. One woman from a small, religiously based NGO supporting sustainable agriculture explained that they had been watching for any changes to the maize plants ever since they learned of discussions about the presence of transgenic maize in Mexico. At first, when they encountered a deformed plant, she told the farmers, "well, this could be the result of transgenes, but we still can't confirm it . . . just keep an eye on your fields, watch your plants." She also encouraged the farmers to try to improve their soils, as she thought the deformed plants could also be a result of nutrient depletion. But, as she told me, "in the meantime, we were learning more about transgenic maize, which is why we became more concerned that perhaps the deformities were a result of the contamination" (agroecology promoter, Oaxaca, March 27, 2006). Most of the petate participants who I interviewed maintained that the hypothesis of a link was merely tentative and not scientifically established. They encouraged attention to malformations because, whatever the cause, it was harmful to the maize harvest. Nevertheless, a small number still are convinced that there is a causal connection and advocate this view strongly.

In contrast, maize activists outside the petate are skeptical about claims that transgenes are causing plant deformities. Some point out that many of the same malformations appeared in maize prior to introduction of transgenic plants, and that there is no evidence of a correlation with transgenic contamination. One activist who has worked for many years with sustainable agriculture and indigenous rights groups in Oaxaca was particularly insightful about this topic. He commented that talk about deformities came as part of the process of transforming the defense of maize into a social movement from below, rather than a campaign by professional activists. He said:

Certainly [the movement] began with a group of intellectuals, a group of activists who began to speak out in the press. . . . [But] at this moment, yes, I believe that there already is a social movement of the farmers themselves, not only of the middle class, not only of environmentalists, not only Greenpeace or that type of group, but rather there is a social movement, farmers are worried, including in the mode of the campesino tradition of rebellion. They generate myths and legends. If you cross the Sierra Juárez of Oaxaca, they will tell you horrible and false myths about transgenic maize. They will show you some horrible plants and say that this is the fault of transgenic maize. All of this is false, they are imaginary, but the people create it out of a real worry. This is the expression of a popular social movement. (maize activist, Oaxaca, March 3, 2006)

He went on to convey his opinion that for rural communities struggling to understand GE crops, it is reasonable that they have sought a physical manifestation of the "perversion of the soul of maize." He noted that his and his colleagues' response to a rural person's claims about deformed plants is "not to humiliate him, not to disqualify him, but neither to give it recognition. We don't agree that this is the route to take." To the extent that NGOs in the petate were promoting the idea, he said, "I think it is opportunism. I think it is bad, I don't agree with this attitude." Another activist, an agroecology promoter (Mexico City, December 15,

2005) involved in both grassroots development projects and national policy advocacy, pointed out that many malformations were well-known plant diseases, often caused by fungus, and that it was "unscientific" for activists to convey these ideas without solid evidence. While understanding that rural people might reasonably develop such a belief, she thought it was inappropriate for NGOs to perpetuate it as though it were true.

Some of the study coordinators readily acknowledge that a scientific audience would not accept their research, but say that such acceptance is not their major concern. As one of the petate study coordinators (October 20, 2005) put it:

We can't say that transgenes are provoking malformations or disease because to be able to publish that you have to do a complete scientific study and it costs a lot of money, and we don't have a laboratory, nor do we have the credibility. [I] can't publish an article that would be recognized in the world of modern and scientific biotechnology.

He went on to say that even if they cannot publish their findings, they can publicize their concerns about threats to native maize. In other words, even if the activists lack the resources and scientific specialization to publish a scientific article on transgene flow in native maize and its effects, they aim to sound alarms about the introduction of GE maize by spreading news of their speculative findings.

A second round of data collection focused on deformed plants, seeking to determine whether they contained transgenic material. Rural communities across Mexico asked the NGOs that coordinated the original study to test the malformed plants that they were finding in their fields. The NGOs involved hoped that the process of sampling and detection would facilitate a deeper process of movement mobilization. As a key petate study coordinator (Mexico City, December 9, 2005) in this research process explained, after the first study,

We reflected on it and said, well, given the extent [of the contamination], we don't have to keep doing testing. We have to defend maize completely, based on the idea that the risk is [already] here. But afterward, people kept asking and asking, and so, to the people who strongly insisted, we said all right, if the tests are used by the people to generate processes [to defend maize] in their communities, we will continue doing the tests for the people who really benefit from it.

They analyzed 173 malformed plants taken from four regions of Mexico. Of these, 17 plants were found to contain transgenes (AJAGI et al. 2005).

Even though the results did not suggest a correlation, petate members continued to claim that there might be some causal link to GE contamination. For example, the NGOs coordinating the study told the press about deformities, maintaining that there was some possible connection to transgenic contamination. A small number of rural activists in Oaxaca are solidly convinced to this day that there is a link, and actively promote the idea in rural communities. They do this, in part, because it helps them explain genetic engineering to people who find it difficult to understand. Others are not so convinced, although they believe it would make things easier if farmers no longer had to rely on laboratory analysis to detect contamination. One agroecology promoter (Oaxaca, April 4, 2006) in the Mixteca region told me that if they knew that transgenes caused deformities, they could tell people, "Look, this malformation indicates to us that there is contamination." He added, it "would help us a lot, but we also hope that it doesn't happen."

Those who felt strongly that GE maize was causing the malformations suggested that the tests themselves may not be capable of picking up signs of transgenes in later generations of plants. I heard this claim made repeatedly by activists who were increasingly skeptical of the utility of the existing diagnostic tools. They had come to believe that transgenic sequences may split up during recombination and therefore might not be recognized by the molecular tests that they use. Several members of the activist network noted that they may be getting false negatives—that is, plants that test negative may still contain parts of transgenic sequences.[8]

The agricultural scientists I interviewed thought that the idea that transgenic contamination might lead to maize deformities was dubious at best. Yet they each offered completely different explanations for why the malformations might be occurring now. One government maize researcher (Oaxaca, March 28, 2006) insisted that he had *never* seen a single malformation in all his years of work, and contended that a link between transgenes and malformations was implausible. He believed that this was disinformation spread by activists to incite fear about GE crops. An agronomist (Oaxaca, March 22, 2006) indicated to me that such malformations had *always* occurred, and that producers were simply paying more attention since the discovery of GE maize contamination. Offering yet another explanation, a biologist (Mexico City, November 25, 2005) confidently stated that the malformations that were being reported were the result of "inbreeding" that caused recessive genes to be expressed. One government maize researcher (Oaxaca, February 6, 2006) thought the link between malformations and transgenic maize could be worth investigating, saying it might be possible to create disease susceptibility if the transgenic plant is not adapted to the region. Yet he indicated that a higher priority would be to identify more likely causes—which could include fungus, bacteria, or even a physiological change due to outcrossing with wild relatives—and find a solution to the malformations that farmers were observing, since such diseases reduce the amount of maize harvested. He suspected that what they were seeing could be the result of a fungus that is frequently introduced when sorghum is cultivated.

Given the contradictory alternative explanations from scientists for the appearance of deformed maize plants, it is not unreasonable that members of the petate continued to advance their own hypothesis. Remaining convinced that images of deformed plants raise compelling concerns about GE maize, rural activists extensively photographed the plants found with malformations. Photos of malformed plants have been put to work in efforts to teach campesinos about biotechnology. One agroecology promoter (Oaxaca, March 27, 2006), herself the daughter of a campesino, told me:

Yes, it is difficult to explain [genetic engineering], but it is easy once we present the photographs of deformed maize plants. We can't completely explain the transformation that is done to the seed, with all the manipulation of genes, but what we can explain is the result, because now we have it, now we bring the photographs and we tell them, look what is happening in that region, or look what is happening in your village. Maybe it isn't your field, but it is that of your neighbor [and so] you have to take measures [to prevent it].

In early 2005, activists made large posters of some of the photographs and staged a demonstration in the House of Representatives of the state of Oaxaca. Maize producers also brought deformed plants, which they held in protest. The demonstration's aim was to urge the state lawmakers to approve a state law that would protect maize biodiversity. The LBOGM had been passed at the federal level in December 2004, and activists hoped that the state would be able to create stronger regulations than those set out by the federal law. Participants in this demonstration did not feel that their protest had any effect, however. This further solidified the conclusion that policy advocacy was not worthwhile. As one rural indigenous rights activist (Oaxaca, March 6, 2006) put it, the demonstration in Oaxaca "didn't have the reverberations that we hoped, and so, well then, we're going to have to continue working in our communities."

Lessons from the Defense of Maize

This case challenges simple binaries that contrast "global science and technology" with "local/indigenous cultural resistance." Maize activists, both within the petate and more generally, are oriented toward global scientific debates and risk discourse. When the petate made claims about maize malformations, it did not present such assertions as indigenous or local knowledge but rather scientific knowledge that scientists had not yet recognized. Indeed, the petate reconstructed the technologies of maize monitoring to meet its self-defined needs for information and conscious-ness-raising. Nevertheless, because they were extremely distrustful of sci-entific and political institutions—perhaps with good reason—members of the petate did not defer to the judgment of experts, leading them to stand by claims that others found to be dubious. In contrast, maize activists who expressed greater trust in scientific experts and at least some institu-tions of government were more careful to avoid making claims that were not widely credible. Maize activists both inside and outside the petate were equally committed to generating processes of community autonomy, grassroots development, and agroecology. Most rejected neoliberal agri-cultural and trade policies, and wanted a ban on GE maize. Where they parted ways was in their assessment of what could be gained through pressure on the federal government and their degree of deference to the authority of scientists.

The case of the network in defense of maize offers several insights about activist-led environmental monitoring and its role in broader pro-cesses of social movement mobilization. First, activist-led monitoring is not just about generating facts; it also is a process of consciousness-rais-ing and solidarity formation. The activists who coordinated the study were conscious of this, strategically approaching the research as an op-portunity to recruit more participants to the struggle against GE maize. Strategies of environmental monitoring may initially be conceived as a necessary response to knowledge gaps or perceived flaws in official risk assessments. Yet knowledge about pollution is not the only or necessar-ily the most significant outcome of monitoring projects. In the petate's case, participation in the research project became the basis for a new set of solidarities among NGOs, activist groups, and widely dispersed rural communities.

Second, this case shows that the expected audience for the findings of activist-led monitoring projects affects the kinds of claims that activists make. The target of monitoring efforts is not always scientific or politi-cal authorities. Certainly, in the early stages of the research effort, the study coordinators conceived of the project as a way to challenge offi-cial contentions that maize contamination was minimal or nonexistent.

As it became evident that Mexican regulatory officials were not swayed by the activists' discoveries and protests, however, the petate stopped viewing the state as a central target of its efforts, and solidified its commitment to cultural projects within rural communities. At the same time, rural community members continued to seek factual information about the presence of transgenes in maize. The monitoring activities therefore continued, but with a self-conscious emphasis on providing information that empowers rural communities rather than engaging in scientific credibility contests. Although organizers were aware, for example, that the claim about malformations would be dismissed by a scientific audience, they knew that the knowledge they gathered would be useful to an audience of maize producers and would publicize their grievances more generally.

Third, the case indicates that monitoring efforts using scientific detection tools may not always be helpful in making "invisible risks" visible to the public. Despite the participatory nature of the research project, the scientific monitoring of transgenes at the molecular level did not necessarily make genetic contamination any more comprehensible to rural communities. Thus, when the first round of genetic tests raised the possibility that transgenic contamination might be causing visible malformations, this hypothesis received particular attention by petate members. Not only did pointing to deformed plants resonate with the idea that biotech can have unpredictable effects but it also seemed to speak to folk beliefs that maize has a soul that can become corrupted. For many, the malformations themselves, whatever the cause, were crucial to publicize because they signaled problems in the maize population and a need for concerted action to protect native maize cultivation. Some organizers believed there was a causal connection between transgenes and malformations, suggesting that visible deformities made monitoring for transgenes possible without having to rely on expensive tests and distant experts. For those involved in the petate, visible deformities, which could be photographed and easily recognized, were much more compelling than data resulting from genetic tests.

Fourth, the disagreement among activists regarding deformed maize plants reveals that in an activist network in which different segments or factions choose different targets (such as the state, culture, and other institutions), conflicts may emerge over the kinds of knowledge claims that are appropriate to make. Disagreements over the credibility of maize monitoring data are inextricably linked to collective identities that are forged in relation to sources of political power. All the social movement organizations involved in the maize movement in Mexico expressed their grievances using risk statements and oriented themselves toward scientific debates, but they did so using different styles, with different objectives, and with different rationales for what makes an assertion credible. Notably, members of the petate were more skeptical than other maize activists about the gains that could be made through existing institutions of government and tended to distrust any claims made by government scientists, whereas maize activists with frequent contact with regulatory officials tended to share those officials' ideas about what constitutes credible scientific knowledge.

Finally, continuing on the previous point, environmental monitoring is unlikely to have an impact on regulation or public policy when there is deep distrust and antagonism between challengers and the state. The present case contrasts with some prominent examples of civil society research in which activists developed relationships with academic and regulatory scientists, eventually having an influence on the regulation of chemicals, or leading to regulatory or legal decisions in their favor.[9] It is possible to imagine that such an outcome might have happened for the petate, had there been some mutual trust between the activists and someone within the government's environmental agencies or a concerned scientist willing to publicly collaborate with them. As mentioned above, this did occur in at least one instance, when rural communities working with one organization shared maize samples with government scientists at the INE, and felt secure in the results reported that no transgenes were discovered. That NGO has been involved in policy discussions at the federal level despite strong opposition to current policies. Organizations in the petate, however, did not enjoy access to government officials, nor did their experiences give them reason to trust the government to deal with the issue of maize contamination honestly.

The movement in defense of maize confronts the dominant risk discourse with its own risk claims and scientific expertise. Yet even as the controversy over GE maize became highly scientized, maize activists reconstructed the methods of scientific analysis—first, the expert assessment, as seen in the previous chapter, and second, environmental monitoring—in order to build a broader movement in support of alternatives to industrial agriculture. In the next chapter, molecular testing is again an essential part of antibiotech activism, but with very different issues at stake. In Canada, where biotechnology companies claim patent rights on transgenes, genes out of place pose problems for farmers who practice seed saving. There, struggles to defend alternative pathways in agriculture play out not in the scientific field, but in the courtroom, where the social significance of genes out of place is determined through legal precedent and judicial rulings. Whereas maize activists found opportunities to reconstruct scientific methods to serve broader struggles for indigenous autonomy and agricultural reforms, antibiotech activists have found the judicial system to be far less receptive to questions about the social implications of transgenic crops.

5

Patents on Out-of-Place Genes

Throughout this book, I have argued that conflicts over GE crops are not merely disagreements about the scientific evidence of risk. Rather, they are disputes about the social order: *What kind of agriculture do we want?* Nowhere is this more obvious than in contestations over intellectual property rights. Social scientists have been attentive to the various court rulings that, over the past thirty years, have established the legal right to patent transgenes and, in some countries, whole transgenic organisms. Kloppenburg ([1988] 2005) and Scott Prudham (2007), among others (e.g., Mascarenhas and Busch 2006), have convincingly argued that patent rights for biological materials—which some critics refer to as "life patents"—enable capitalist firms not only to treat those materials as commodities to be bought and sold but also overcome the self-reproducing character of living things, and thus put a halt to farmers' traditional seed-saving practices. As Prudham (2007, 413) puts it, "Without stringent protections of exclusive rights, pollen drifts, mice reproduce, and so on, potentially undermining the realization of invested capital."

In March 2011, the Public Patent Foundation (PUBPAT), a nonprofit legal services organization representing a diverse group of US and Canadian farmers, seed producers, and organic agriculture organizations, filed suit against Monsanto, the world's largest seed company. This complaint was significant, as it marked a new approach in a series of struggles to establish the legal status of genes out of place. The plaintiffs sought no compensation, only an assurance that Monsanto would not be allowed to sue them for patent infringement if their crops were ever contaminated with transgenic material that is patented by the company. Monsanto, which controls over a fifth of the world's proprietary seed market, and produces the seeds planted on 80 to 90 percent of US corn and soybean fields, is known for its aggressive use of patent rights in lawsuits against farmers in North America (Hubbard and Farmer to Farmer Campaign on Genetic

Engineering 2009; Center for Food Safety 2005). One previous legal case, in particular, drew international attention to the possibility that Monsanto might sue farmers who never intended to grow GE crops or take advantage of their patented traits. In 2004, the Canadian Supreme Court ruled in favor of Monsanto Canada against Percy Schmeiser, a farmer from the province of Saskatchewan. Schmeiser had saved and planted seeds from his own canola harvest, although he had been informed that his canola contained genetic material that was patented by the biotechnology company. The ruling indicated that knowingly planting seeds that contain patented genes—even if those genes arrived in a field through pollen drift, wind-borne seeds, or other accidental means—constitutes patent infringement.

I examine the case of Monsanto versus Schmeiser in this chapter, and discuss its implications for other efforts to use the courts to settle disputes about genes out of place and intellectual property. There have been numerous academic and popular analyses of the Schmeiser case, but here I make several new observations.[1] First, the case demonstrates that even debates about property rights and seed saving—eminently political topics—have become scientized, making it difficult for farmers, activists, and other nonscientists to have a substantial role in shaping public policy. In Canada, as in the United States, there are high "expertise barriers" in the patent domain that limit public scrutiny and challenges from outsiders (Parthasarathy 2010, 2011). Despite pressure for the legislature to debate the social consequences of biotechnology patents, the topic has been left primarily to the patent offices and judicial system to decide. In *Monsanto v. Schmeiser*, the courts declined an opportunity to open gene patents and the regulation of GE crops to public scrutiny and debate. Much as in the case of GE maize in Mexico, the technical aspects of monitoring transgenes in a plant population became the central focus. Indeed, the legal dispute did not primarily revolve around questions of seed-saving rights and the social consequences of patents but rather around questions of proper research methods and an accurate understanding of how genes affect plant development.

Second, the Schmeiser case influenced the way that antibiotech activists around the world conceive of the threat from genes out of place. The Supreme Court ruling is widely understood by antibiotech activists to demonstrate that transgene flow can lead to charges of patent infringement—a worry that I often heard repeated by Mexican maize activists, and is reflected in the PUBPAT demands. Finally, the case offers a revealing example of legal mobilization in defense of seed saving and against GE crops. Even though neither Schmeiser nor antibiotech activists initiated the legal conflict with Monsanto, the case became an important arena for challenging the system of intellectual property rights for plant genetic resources. It is thus a productive site for examining how social movements use science in the courtroom as part of efforts to produce social and technological change.

It is impossible to know exactly why Schmeiser took the actions he did, or what he actually knew, at key moments, about the existence of patented genes in his canola seeds. There are many critics of Schmeiser who doubt his credibility. Advocates of biotechnology who I interviewed were uniformly critical of Schmeiser, generally believing him to be dishonest about his dealings with GE canola, and seeing him as lacking trustworthiness because of the financial support they claim he receives from Greenpeace or other antibiotech organizations. On its Web site, Monsanto says that Schmeiser is "simply a patent infringer who knows how to tell a good story." My aim here is not to uncover what Schmeiser *really* did, although I do find that the evidence against him was not particularly strong. Rather, I am concerned with the ways that the judges dealt with scientific evidence in arriving at their rulings and the role of this case in the broader antibiotech movement. Whatever Schmeiser's intentions, his conflict with Monsanto not only set a legal precedent regarding the legal status of genes out of place; it also drew public attention to the social implications of transgene flow.

I first discuss the social context in which the accusations against Schmeiser became a heated public controversy. Specifically, I describe the rapid growth of an intellectual property regime for plant genetic material and a corresponding mobilization in defense of seed saving. I then consider the role of legal mobilization—social movement struggles enacted through judicial institutions—in opening up science and technology to broad public scrutiny, and supporting alternatives to the dominant trajectory of technological change. Next, I turn to the details of the Schmeiser case, beginning with some background on his farming practices, and followed by an analysis of the trial and rulings against Schmeiser. Here, I focus on the role of scientific knowledge—particularly knowledge about transgene flow—in the justices' decisions. I consider what this case indicates about how legal rulings establish the social significance of genes out of place, and how legal mobilization figures into broader processes of public engagement in science and technology.

Biotechnology Patent Conflicts

Since the origins of agriculture, humans have saved seeds from their harvest to plant the following season. This practice remains common in peasant agricultural systems in much of the Global South, and is also typical among producers of certain crops in the industrialized agricultural systems of the Global North. Scientific developments such as plant hybridization in the 1930s and legal concepts such as PBR, developed in the 1970s, were major steps toward transforming self-reproducing biological material into commodities and creating markets for seeds where they did not previously exist (Kloppenburg [1988] 2005). Intellectual property rights and genetic engineering to render plants sterile are the most recent forms of controlling access to genetic resources.

Since the 1980s, the international governance of plant genetic resources has shifted from a common heritage system, which treated plant genetic resources as a commons, to a system that emphasizes sovereign and private property rights (Raustiala and Victor 2004, 284). Biotechnology patents, first granted in the 1980s, recognize GE plants and organisms as inventions. In 1980, a landmark US Supreme Court case, *Diamond v. Chakrabarty*, determined that a bacterium created by Ananda Mohan Chakrabarty, a genetic engineer working for General Electric, could be patented under existing US law. This ruling facilitated the rapid emergence and development of the biotechnology or "life sciences" industry. By the mid-1990s, intellectual property rules were built into international trade agreements such as NAFTA and the WTO, requiring countries that became signatories to those agreements to treat genetic material as intellectual property. There is vast "regime complex" today of international agreements and institutions addressing the rights of plant breeders and genetic engineers to claim the exclusive privilege of reproducing and selling the plants and seeds they modify (Raustiala and Victor 2004). That is, in addition to national laws, there are several different international arrangements governing plant genetic resources, including: the International Convention for the Protection of New Varieties of Plants, which governs property rights over intentionally bred plant varieties; the International Treaty on Plant Genetic Resources, negotiated through the UN Food and Agriculture Organization; the WTO's Agreement on Trade-Related Aspects of Intellectual Property Rights; and the CBD.

In Canada, the Patent Office granted the first patent on a GE microbe in 1982.[2] In addition to patents on GE material, new policies regarding intellectual property rights for seeds developed through conventional

breeding methods went into effect in 1990, when Canada adopted a system of PBR. Canada's PBR system gives limited monopoly rights to plant
breeders for the sale, propagation, and use of the seeds they develop, for a
period of eighteen years. Under this law, farmers are allowed to save and
reuse seeds from plants they grow from seeds covered by PBR, but they
are not allowed to sell those seeds.[3] The Canadian patent system does not
include provisions for the public to object to a patent on religious, ethical,
or social grounds.[4] Michelle Swenarchuk, an attorney and environmental
activist at the Canadian Environmental Law Association, notes a lack of
public debate about "life patents":

> Canada urgently needs now what should have occurred before life patents were
> approved twenty years ago, a full public debate, not only amongst governments
> and legal, scientific and ethical experts, but with appropriate consultation with
> all Canadians. The debate should include a credible examination of the totality of
> impacts of life patents and result in law reform that re-balances the law to better
> accord with the original social purpose of the patent system and contemporary
> Canadian values. (Swenarchuk and Canadian Environmental Law Association
> 2003)

The Canadian government made an attempt to encourage public participation in the patent system, but was criticized by activist groups, which
did not view it as a truly open process. In 1999, the Canadian government created the Canadian Biotechnology Advisory Committee (CBAC),
an expert panel tasked with providing advice and facilitating public dialogue on biotechnology policy issues. One of CBAC's major projects was
a consultation process to solicit public input on intellectual property and
the patenting of higher life-forms. Greenpeace, the Council of Canadians,
and the *Ram's Horn* (a monthly journal on agricultural issues) called for a
boycott of the process, "viewing the workshops as a potential 'participation trap.' That is, the NGOs suspected that CBAC could use participation in the workshops as a basis upon which to legitimate its recommendations, whatever these might be." One analysis of this episode concluded
that the activists "identified the essential problem of purely discursive
democracy under conditions of unequal power relations" (Prudham and
Morris 2006, 162).[5] In other words, they recognized that even if they
participated, more powerful actors in the biotechnology and agricultural
industries would have a far greater influence over the outcome.

The vast majority of farmers in North America have accommodated
themselves to the intellectual property system, adopting GE seeds and
purchasing them anew every year from seed dealers.[6] A small but influential set of opponents of this system still exist, though. As indicated

in earlier chapters, research and activism surrounding access to plant genetic resources was essential to the development of the antibiotech movement since the 1970s. Throughout the 1970s and 1980s, activists in the Global North and Global South demanded a system in which all plant genetic resources, including commercial cultivars developed by plant breeders, are considered the common heritage of humankind and therefore not open to privatization (Kloppenburg [1988] 2005, 173).[7] Today, ETC Group and Genetic Resources Action International (GRAIN), prominent organizations that raise questions about how corporate control of genetic resources might threaten global biodiversity, are active in Canada. ETC Group has an Ottawa office, and Devlin Kuyek (2004, 2002, 2007a, 2007b), a Montreal-based researcher for GRAIN, monitors and writes extensively about seed politics as well as the biotechnology industry in Canada.

In addition to genetic resources advocacy groups, a combination of Canadian farmer organizations, environmental organizations, and consumer advocacy groups add to the voices of dissent against biotechnology patents and corporate concentration in agriculture. Some take an ethical or religious position against the treatment of living things or living material as inventions. The Canadian Council of Churches, for example, works to stimulate ethical reflection about the implications of awarding patents on DNA, plants, and animals, asking whether patents on human organs and the "ownership of life" are far behind. Another source of opposition to biotechnology patents are farmers and farming advocacy organizations that defend the right of farmers to save and reuse seed. And Canada's NFU has opposed not only patents on GE crops but also other public policies that limit the freedom of farmers to save seeds. For instance, in 2004 and 2005, the NFU fought hard to defeat proposed changes to Canadian law that aimed to expand PBR beyond what was established in the 1990 law. The proposed changes would have put increased restrictions on farmers' rights to save and reuse seed on their own farms.

ETC Group, GRAIN, the Canadian Council of Churches, and the NFU are joined by numerous other Canadian organizations that oppose the use of patents to restrict farmers' seed saving—many of which are members of the Canadian Biotechnology Action Network (CBAN). CBAN has had a notable impact on the biotechnology industry in Canada. The network of antibiotech organizations successfully opposed the commercialization of recombinant bovine growth hormone (a drug produced through genetic engineering that increases milk production in dairy cows) and pressured Monsanto to withdraw its application to sell GE wheat seeds.[8]

CBAN has also maintained pressure on the Canadian government regarding genetic use restriction technologies (GURTs), dubbed "Terminator" by antibiotech activists. GURTs are genetic manipulations that make plants produce sterile seeds (seeds that will not grow). Like intellectual property rights, GURTs are intended to require farmers to purchase new seeds for planting every year. Since 1998, ETC Group has led a global campaign to ban GURTs. In 1999, Monsanto and AstraZeneca (now Syngenta) pledged not to commercialize GURTs, but interest in the technology remains strong. Since 2000, an expert group convened under the international CBD has recommended that governments not allow the testing and commercialization of GURTs. In 2005, however, the Canadian government sought to change the CBD recommendations to allow for the field testing and commercialization of GURTs, pending science-based risk assessments. In response, Canadian-based civil society organizations, including ETC Group and the NFU, among others, initiated a Ban Terminator Campaign. The campaign grew to include over five hundred organizations worldwide, arguing that GURTs would impose billions of dollars of extra seed costs on farmers worldwide (Ban Terminator Campaign 2006). In 2006, the CBD voted to maintain a moratorium on GURTs. Antibiotech activists have also sought to ban GURTs within Canada, trying on repeated occasions to pass legislation that would place a moratorium on the technology.

Social Movements and the Law

In Canada, social movement struggles over the patenting of genetic material and life-forms have repeatedly taken place in the judicial system, but research on legal mobilization indicates that social movement activity in the courts often yields ambiguous results. The outcomes of legal mobilization—using the law as part of struggles for social change—are difficult to predict and depend largely on social context (McCann 2006). When legal advocacy works to produce social change, it is usually in combination with other tactics, "such as public demonstrations, legislative lobbying, collective bargaining, electoral mobilization, and media publicity" (ibid., 31). While some argue that, by and large, even successful legal challenges do not help social movements (Rosenburg 1991), a recent review of the literature suggests:

Legal mobilization tactics do not inherently empower or disempower citizens. Legal institutions and norms tend to be Janus-faced, at once securing the status quo of hierarchical power while sometimes providing limited opportunities for episodic challenges to and transformations in that reigning order. How law matters depends on the complex, often changing dynamics of the context in which struggles occur. (McCann 2006, 35)

Engaging in legal mobilization can have a wide range of impacts on social movements themselves. During the early phases of the formation of a social movement, law can either aid in mobilization or thwart movement development. On the one hand, legal action can aid in the process of consciousness raising that makes it possible for aggrieved citizens to perceive their problems as a matter of legal rights (Scheingold 1974). This can lead to more defiant mass protest, which in combination with litigation can produce social change. On the other hand, "legal action often fails as a resource for expanding social movement activism . . . largely owing to the absence of favorable social conditions" (McCann 2006, 28). Legal tactics can discourage or thwart collective action, for instance, by diverting resources toward lawyers rather than grassroots mobilization or other forms of political organizing. Framing issues in terms of the law may also lead a movement to lose its more critical, transformative edge.

As far as the contentious politics surrounding new technologies such as GE crops are concerned, Jasanoff, a prominent scholar of science in the law, observes that the courts have an important role in shaping how a society interprets and responds to technology. When the public is concerned about the social and cultural dimensions of science and technology, judges are often the ones to decide whether those views will be respected, "even if they cut against the managerial preferences of the nation's scientific and technological elite" (Jasanoff 1995, 13). Again, though, legal institutions are double-edged. The courts may make decisions that stimulate the public examination of previously unvoiced political and ethical issues, or they may mute conflicts by affirming scientistic regulatory structures and constraining public debate—an outcome seen in a series of cases involving transgenic organisms in the United States. A recent study of cases involving controversial technologies indicates that "judges, through various private law principles, support and legitimize novel technologies" (Chandler 2007, 348).

In the early 1980s, Jeremy Rifkin, one of the most prominent US critics of biotechnology, successfully sued the EPA to stop field tests of a GE microbe designed to inhibit the formation of frost on fruits and vegetables. The case brought public attention to the shortcomings of the then-nascent regulatory process. This was followed by a series of lawsuits brought by Rifkin's organization, the Foundation on Economic Trends, challenging the development of biotechnology. In most of these cases, however,

Rifkin's side lost—an outcome that Jasanoff (1995, 158) has suggested "attests to the limitations of legal proceedings as a forum for framing and conducting meaningful technology assessment." Although these lawsuits aimed to generate wide-ranging public discussion about the potential impacts of biotechnology, the US courts "left intact a risk-based, heavily bureaucratic approach to regulating biotechnology that did not in the end encourage open-ended moral or ethical questions" (ibid.).

On the other hand, there is at least one example in which a court case stimulated ethical debate about biotechnology and led to a ruling that challenged the dominant trends. In 1993, Canadian patent examiners rejected Harvard College's request for a patent on a genetically modified mouse, saying only genes, but not the whole organism, could be patented. Harvard appealed the decision, ultimately reaching the Supreme Court of Canada. Eleven religious, environmental, and animal welfare organizations served as interveners in support of the patents commissioner, including the Canadian Council of Churches, Greenpeace Canada, the Canadian Institute for Environmental Law and Policy, ETC Group, and the International Fund for Animal Welfare (Furlanetto 2003). As the range of organizations serving as interveners indicates, the legal battle over the so-called Harvard mouse or "oncomouse" became a focal point for critics from a wide range of perspectives, from animal rights groups to Christian organizations. Ultimately, in December 2002, the Supreme Court upheld the decision to reject the patent application, concluding that higher life-forms cannot be patented. The Canadian Supreme Court's ruling broke with precedents set by the United States, Japan, and the European Patent Office, which allow patenting on higher life-forms. In Canada, therefore, intellectual property protections for whole plants and seeds are available only under the Plant Breeders' Rights Act, which offers less expansive rights than a patent.[9]

Beyond the rulings themselves, one potential outcome of legal mobilization surrounding issues of science and technology can be an increased questioning of scientific authority and the empowerment of citizens to assert their own expert knowledge. Social movement struggles in the judicial arena do not yield predictable outcomes, yet as Jasanoff (1995) and others indicate, the courtroom offers an important venue for what Hess (2007) calls epistemic modernization. In legal contests, scientific knowledge and expert authority are opened up to scrutiny by diverse publics. As other scholars have noted, the use of science and expert witnesses as trial evidence is often fraught with contentious credibility contests (Lynch and Cole 2005; Caudill and LaRue 2006). Jasanoff (1995, 54) finds that in the adversarial process, "all the factors that go into establishing [an expert] witness's credibility—not only knowledge but also social and cultural factors such as demeanor, personality, interests and rhetorical skills—are simultaneously open to attack." She suggests that this does not necessarily imply the distortion of scientific knowledge but rather reveals already-existing inconsistencies and professional disagreements within and between scientific disciplines. The adversarial process exposes scientific experts to public scrutiny in a way that does not usually occur in other settings. Thus, when activists challenge new technology in the courtroom, these struggles have implications not only for how society deals with the technology but also for the authority of scientific knowledge.

Furthermore, legal mobilization may lead to new understandings of *whose* knowledge counts as relevant expertise. Jasanoff (ibid., 19) opines that many lawsuits raise the questions: "Whose knowledge should count as valid science, according to what criteria, and as applied by whom? When should lay understandings of phenomena take precedence over expert claims to superior knowledge?" As in the other possible outcomes of legal mobilization, it is difficult to predict what will result from such lawsuits. Judges may decide, for instance, that a layperson's knowledge is superior to that of university researchers in a particular case, or they may choose to treat only credentialed experts as authorities.

The case I discuss in this chapter highlights a kind of legal mobilization that has not received attention from scholars of social movements and the law. It involves a dispute over intellectual property, in which antibiotech activists reacted strongly against the legal machinations of a powerful biotechnology company, Monsanto. In 2005, the Center for Food Safety (CFS), a US-based environmental advocacy organization, published a report documenting Monsanto's campaign of investigations and lawsuits against farmers who violate Monsanto's Technology Use Agreement or use patented genes without signing an agreement (Center for Food Safety 2005). The CFS and others argue that Monsanto is abusing its intellectual property rights in order to gain monopoly control of the seed market.[10] According to the report, Monsanto "has built a department of 75 employees and set aside an annual budget of $10 million for the sole purpose of investigating and prosecuting farmers for patent infringement" (Center for Food Safety 2005, 23). An update to that report found that as of October 2007, Monsanto had filed 112 lawsuits against farmers, and investigates hundreds of farmers each year (Center for Food Safety 2007). It appears that many of these investigations result in out-of-court settlements. The CFS used Monsanto documents to estimate that the company had pursued between two thousand and forty-five hundred "seed piracy matters" against US farmers as of June 2006, resulting in settlement payments totaling somewhere between $85 million and $160 million (ibid.).

In many of the cases that the CFS documented, the farmers purchased seeds from Monsanto and then saved seed to plant, perhaps not aware of or understanding the Technology Use Agreement, or acting in defiance of the patent system. In some instances, those accused of patent infringement by Monsanto say that the company falsely targeted them, but nevertheless intimidated them into paying a settlement (Barlett and Steele 2008; Center for Food Safety 2005). In one case, in 2004, acting on an anonymous tip, Monsanto accused Indiana soybean farmer David Runyon of patent infringement. Runyon had only ever planted publicly developed, nonpatented seed. But when Runyon himself tested his soybeans, he discovered that Monsanto's patented material was present, as a result of contamination. Runyon was able to use an Indiana farmer protection law to shield himself from Monsanto (Adams 2009). Eventually, Monsanto stopped pursuing Runyon, but publicly stated that he is ineligible to purchase its seeds unless he "cooperates" (Monsanto 2011a).[11]

Intellectual property conflicts are rather unique compared to the forms of legal confrontation that have so far been analyzed by scholars of social movements and the law. The closest parallel to be found in the sociological literature is studies of criminal prosecutions of activists. While activists may take the role of plaintiff (in class action lawsuits, for example, or famous cases such as *Brown v. Board of Education*), activists are frequently in the defensive role, typically as the outcome of civil disobedience. The legal system can be used to harass activists and hinder protest movements, as was the situation in the US civil rights movement. In some circumstances, however, legal defense proceedings can generate forums for activists to express their political viewpoints and make moral arguments for their acquittal. This was often true during prosecutions of Vietnam War protesters in the United States (Barkan 1985, 2006). Patent disputes are not criminal proceedings, but they do harass farmers who resist the dominant system for distributing genetic resources. At the same time, they contain the potential to open up a forum for moral claims about patenting and seed saving.

Schmeiser stood up to Monsanto in court despite the enormous resources required to do so and then went on to become a world-famous critic of biotechnology. To defend himself, Schmeiser not only argued that the patent laws were flawed, with unacceptable social consequences, but also made a scientific argument about transgene flow, the contamination of the seed supply, and the shortcomings of the research methods used by Monsanto's investigators. His case is therefore important—and unique—in terms of examining the circumstances in which patent lawsuits might offer a forum for social protest.

Schmeiser's Story

I interviewed Schmeiser at his home in July 2006. At the time, Schmeiser had been farming for half a century in Bruno, Saskatchewan, on land that was once farmed by his father. The western prairie province of Saskatchewan at that time of year is covered in vast fields of still-green wheat and oats, bright yellow canola, and cool violet flax. Country grid roads, built before Canada adopted the metric system, divide the large, flat province neatly into one-by-two-mile blocks. A large proportion of the world's canola is produced here; indeed, Saskatchewan is the top canola-producing province in the world's largest canola-exporting country. In 2010, Saskatchewan farmers grew 7,400,000 acres of canola. A great majority of Saskatchewan canola producers use genetically modified or other herbicide-tolerant varieties. Only 7 percent of canola planted in Canada in 2010 was not transgenic.[12]

Like most other grain farmers in the province, Schmeiser typically used chemical fertilizers, herbicides, and pesticides when necessary to produce a good crop of oats, wheat, peas, and canola. While dependent on chemical inputs, until 1999, Schmeiser said he rarely found it necessary to purchase seeds. Indeed, he considered himself an expert in seed improvement, purchasing seeds only as a means to introduce beneficial traits into his seed supply. By the 1980s, for instance, he had developed a variety of canola that was resistant to diseases such as blackleg. It is typically recommended that farmers only cultivate canola every four years to avoid vulnerability to plant diseases, but with the varieties he improved, Schmeiser said he was able to grow canola continuously.

As Schmeiser recalled, farmers in the area started cultivating a new kind of canola in 1996, signing Technology Use Agreements with Monsanto, the maker of Roundup herbicide, and paying for permission to use the patented Roundup Ready canola plants. The package of seeds and herbicide offered a new method of weed control. Farmers could simply plant the seeds, let the plants grow, and spray the fields with Roundup to kill all the weeds, leaving a tidy, "clean" canola field. In 1997, only about 4 percent of the canola grown in Canada was Roundup Ready. By 1998, this number had jumped to 23 percent, attesting to the growing popularity of the technology.[13] But Roundup did not change Schmeiser's farming practices. Since he rarely went to the seed dealer, he simply continued to plant his own canola seeds. Beginning in 1997, however, Schmeiser's canola farming practices began to have significant legal consequences.

In 1997, Schmeiser sprayed Roundup herbicide on the weeds growing around the power lines running through his property (*Monsanto v. Schmeiser* 2001, par. 38–39). Stray canola plants, one of the major weeds in Canadian agriculture, were among the weeds that were sprayed that year, and they did not die. As Schmeiser explained it, he thought that perhaps the plants had started naturally becoming tolerant of the herbicide. He therefore conducted a test in one of his fields, spraying about three acres of his canola with Roundup. By his estimates, about 60 percent of the plants, mostly those closest to the edge of the field, survived the spraying and continued to grow. It is not entirely clear from the evidence presented by Monsanto and Schmeiser where the Roundup-resistant plants came from. Schmeiser's hired hand testified that in 1996, there had been strong winds during the harvesting of canola on Schmeiser's and neighboring farms, which meant that canola from other farms may have ended up in the seeds that Schmeiser saved to plant the next year (*Monsanto v. Schmeiser* 2000, par. 22).[14]

At harvest time, in 1997, Schmeiser's hired hand harvested canola seeds from the field where Schmeiser had discovered herbicide-tolerant plants, and stored a portion of them in the back of a truck. Schmeiser attested that he harvested canola from the entire field, while his accusers suggested that he saved only the seed that survived the Roundup spray. Schmeiser did not contest that he saved and planted seed from the field in which he found Roundup-resistant canola growing around the edges in 1997. But he denied Monsanto's accusations that he did so with intentions to use the patented Roundup Ready trait. Indeed, he said he mixed the seed from that field with other seeds, thereby making it counterproductive to use Roundup as a weed control on his fields (ibid., 2000, par. 44).

That year, Monsanto hired a private investigator to take samples from the edges of farmers' canola fields in order to determine whether they were growing Roundup Ready crops without a license. Without Schmeiser's knowledge, the investigator took some seedpods from the edge of Schmeiser's field that summer, and the seedpods tested positive for Roundup Ready genes (*Monsanto v. Schmeiser* 2001, par. 41–42). The following spring, Monsanto's private investigator came to Schmeiser and told him that he had evidence that Schmeiser grew Roundup Ready canola the previous season. According to court documents, Schmeiser did not treat this concern with any seriousness (ibid., par. 45). He took the seeds from the field where he found Roundup-resistant plants to a treatment facility (to prepare them with chemicals for planting) and then planted them, as he would with any other seeds saved from his harvest. Monsanto Canada then filed a lawsuit against Schmeiser, accusing him of using its patented Roundup Ready trait without obtaining a license. The company claimed that Schmeiser was using its patented genes without permission and thus was in violation of its intellectual property rights.

To help pay for his legal defense and promote his cause, Schmeiser found eager allies in the antibiotech movement. When I interviewed him in 2006, he maintained that when he was first accused of patent infringement, he knew very little about GE crops and was incredulous that such a charge could result from saving his own seeds. After the Federal Court ruled in favor of Monsanto in 2001, though, Schmeiser became a prominent figure in a transnational network of activists and scientists opposing GE crops and gene patents. When Schmeiser appealed the ruling against him to the Canadian Supreme Court, several Canadian and international NGOs served as interveners on his behalf. These included the NFU, the Sierra Club of Canada, ETC Group, the Council of Canadians, the CFS, and the Research Foundation for Science, Technology, and Ecology. In the midst of his legal troubles, Schmeiser was frequently invited to speak about his experiences at forums around the world, and during these events he drew attention to the negative social consequences of biotechnology patents. To give just one of many examples, he spoke at a civil society forum in Johannesburg, South Africa, during the World Sustainable Development Forum in 2002.[15] According to a report on the forum, Schmeiser talked about the lawsuit against him, and made the points that "patent law overrides farmers' rights" and the climate of surveillance of farmers' activities destroys the social fabric (Third World Network 2002).

Schmeiser's Defense

Schmeiser's attorney made several different arguments in his defense, including contentions against the validity of the patents and the failure of Monsanto to control its invention. Schmeiser's defense asserted that Monsanto should not be able to claim exclusive rights to its invention while at the same time releasing it into the environment, where its self-propagation cannot be controlled. If Monsanto wishes to claim exclusive rights, the defendant maintained, the company should control the spread of the gene (*Monsanto v. Schmeiser* 2000). The defense also argued that a ruling against the farmer would set a precedent for the persecution of all farmers who save and use their own seeds: "If the right of Mr. Schmeiser to save and re-use his seed is taken away from him, any other seed saver is not far behind. Perhaps this is a benefit that Monsanto hoped to achieve by releasing their product into the environment without any control" (ibid., par. 187).

Schmeiser insisted that any transgenic canola in his fields was there because of unintentional pollen and seed drift from neighboring fields, but Monsanto wanted to demonstrate that Schmeiser had intentionally saved, segregated, and cultivated Roundup-resistant canola seeds. Evidence that a high proportion of the farmer's plants were indeed Roundup resistant would strongly support the company's allegations that Schmeiser intentionally segregated and cultivated Roundup-resistant canola. But in 1998, Schmeiser planted 1,030 acres of canola. How could anyone know just how much of the canola growing in that vast area contained the gene?

Monsanto had a number of expert witnesses, including, significantly, agricultural scientist Keith Downey, known as the "Father of Canola" because of his role in developing the varieties of canola grown around the world today. Downey maintained that it was statistically impossible for there to be such high levels of Roundup Ready canola across Schmeiser's fields without intentional segregation and cultivation of the seeds (*Monsanto v. Schmeiser* 2001, par. 112). Accidental contamination alone could not, he argued, produce the findings that were gathered. Schmeiser gained the assistance of two plant scientists at the University of Manitoba, Lyle Friesen and Rene C. Van Acker, who helped to call into question some of Monsanto's scientific evidence. The two scientists tested seed samples from Schmeiser's fields and testified that levels of transgenic material were considerably lower than Monsanto claimed. Furthermore, they offered their expert opinion that the intermingling of GE and conventional canola was widespread, even in the commercial seed supply.[16]

Monsanto had sent private investigators to collect seedpod samples from three places: the edges of Schmeiser's fields in 1997; the mill where he had his seed treated; and all nine of the canola fields that Schmeiser cultivated in 1998. The test results on these three samples were the main evidence against Schmeiser. Schmeiser's lawyer strongly held, supported by statements from Friesen, that none of the tests could be understood as representative of the fields as a whole. The fields were not sampled scientifically, and hence any tests done revealed only "what is in the bags, not what is in the fields" (*Monsanto v. Schmeiser* 2000, par. 55). In addition to the question of the statistical representativeness of the samples, Schmieser's defense raised numerous concerns about the credibility of the samples, pointing out what appeared to be questionable practices by those who collected and tested the seed samples for Monsanto. The handling, for instance, of the 1997 roadside sample seemed suspicious. The seedpods were handled by a long chain of people, and at some point were transferred from their plastic bags, labeled by the same private investigator collecting the samples, into paper envelopes without any identifying information. Questions were raised about the origins of some of those 1997 roadside samples as well, since the description that the investigator offered of the place where they were taken did not match the description of a field in which canola was being cultivated. Schmeiser's defense contended that there also appeared to be manipulation of the samples taken from the seed treatment facility and the samples that were supposed to represent Schmeiser's 1998 canola fields. In both cases, the samples that were tested by Monsanto's experts appeared unusually clean and free of plant debris, compared to what had been taken from Schmeiser's harvest and fields.[17]

In addition to questioning the methods of evidence collection and analysis, the defense also called on the testimony of farmers familiar with the area and local farming practices, and their knowledge contradicted that of Monsanto's expert witnesses. Using this local knowledge, Schmeiser's defense maintained that Roundup Ready seeds and pollen were carried to Schmeiser's land by the wind, accounting for the Roundup Ready seeds detected in samples taken from his fields. For example, Schmeiser contended that Roundup Ready seeds arrived in his fields when they blew off of a passing truck. Monsanto brought an expert witness, a mechanical engineer, to testify on the distance that canola seeds could blow in the wind. His model indicated that a canola seed falling off of a truck would travel only 8.8 meters from the roadway, landing in the ditch (CP Wire 2000). Schmeiser's attorney challenged this opinion, saying that his model did not take into account the wind speeds in Bruno, Saskatchewan. A farmer, Elmer Borstmayer, testified to having driven trucks of Roundup Ready canola seeds covered with a loose tarp past all of Schmeiser's fields in question and that a large quantity of those seeds blew out of his truck. But when I interviewed Schmeiser (Bruno, Saskatchewan, July 20, 2006), he expressed his frustration that the judge did not treat farmers like Borstmayer as expert witnesses:

When I brought in a farmer who was farming for thirty years, an expert in farming, the judge said he's not an expert witness, and that he couldn't sit in the court [to watch the proceedings]. He had to leave, whereas [the experts for] Monsanto, because they were scientists, they're considered experts by the judge. A farmer who farmed for thirty years is not considered an expert. And yet, these scientists knew nothing about farming, neither did the judge. So, talk about being unfair.

Schmeiser felt that the expert witnesses and the judge did not have an understanding of just how windy it is where he farms, and hence how easily canola seeds can spread. On this particular matter, it is evident in the trial judge's ruling that he accepted the mechanical engineer's expert opinion on windblown seed rather than the opinions of farmers.

Judging the Facts

The federal trial judge, Andrew MacKay, believed Monsanto's argument that Schmeiser's fields in 1998 were planted in almost entirely pure Roundup Ready seeds, indicating that Schmeiser had intentionally segregated and planted the seeds. The judge therefore ruled in favor of the biotech company. The ruling stated, "His infringement arises not simply from occasional or limited contamination of his Roundup susceptible canola by plants that are Roundup resistant. He planted his crop for 1998 with seed that he knew or ought to have known was Roundup tolerant" (*Monsanto v. Schmeiser* 2001, par 125). In the judge's opinion, Schmeiser infringed on Monsanto's exclusive patent rights when he planted, harvested, and sold that canola in 1998.

Significant here is that, in this ruling, it did not matter where the Roundup Ready seed came from—contamination or illegal purchase. It only mattered that Schmeiser *should have known* that the seeds contained Monsanto's gene, and he planted them anyway. It also was of no importance to MacKay's decision that Schmeiser did not use the Roundup Ready trait—that is, he did not spray Roundup to clear weeds from the fields of canola. Yet the judge noted in his decision that his ruling was based on a tentative acceptance of the scientific evidence:

Despite questions raised about particular aspects of the sampling and the handling of samples of the defendants' 1998 canola crop . . . the balance of probabilities supports a conclusion that the growing and sale of Roundup tolerant canola by the defendants infringed the exclusive rights of the plaintiffs to use the patented gene and cell. I reach that *tentative* conclusion having also concluded on a balance of probabilities that the samples taken from the borders of nine fields in July 1998 and three samples taken at random from within each field in August 1998 are representative of the entire crop, bearing in mind that all of the nine fields were planted with seed that was saved in 1997 in field number 2, which seed was known to be Roundup tolerant. (*Monsanto v. Schmeiser* 2001, par. 114; emphasis added)

Schmeiser appealed the ruling to the Federal Court of Appeal, which, in a two-to-one decision, supported MacKay's decision. He then appealed to the Supreme Court of Canada. Remarkably, when Schmeiser appealed to the Federal Court of Appeal and, eventually, the Supreme Court, the judges' decisions granted increasing credibility to the scientific evidence against him—in essence, removing the "tentative" qualifier that the trial judge deemed necessary. Indeed, the Supreme Court appeared to ignore the uncertainty of the evidence altogether, asserting as a salient *fact* in its decision that "tests revealed that 95 to 98 percent of his 1,000 acres of canola crop was made up of Roundup Ready plants" (*Monsanto v. Schmeiser* 2004, par. 6). The majority opinion in the Supreme Court ruling notes that the only evidence to contradict that finding is the test of samples of seed that were provided to Friesen by Schmeiser himself. The ruling stated: "A series of independent tests by different experts confirmed that the canola Mr. Schmeiser planted and grew in 1998 was 95 to 98 percent Roundup resistant. Only a grow-out test by Mr. Schmeiser in his yard in 1999 and by Mr. Friesen on samples supplied by Mr. Schmeiser did not support this result" (ibid., par. 64). This statement seems to suggest that the Supreme Court doubted the accuracy of these contradictory findings because of Schmeiser's involvement, although no similar skepticism was shown toward tests done by Monsanto employees.

The Supreme Court ruled in favor of Monsanto (with four out of the nine justices dissenting), but determined that Schmeiser did not owe the company any monetary damages because he made no profits that could be attributed to his use of the patented invention.

After the ruling was announced, Van Acker sent a letter to Beverley McLachlin, the chief justice of the Supreme Court of Canada, expressing his concerns about the ruling's statement of fact with respect to the amount of GE canola in Schmeiser's 1998 fields. In the letter, Van Acker recounted the evidence he presented in court of the much lower and more variable levels of GE canola in Schmeiser's fields—evidence that was discounted by the Supreme Court. He also described his peer-reviewed research with Friesen that revealed widespread contamination of certified pedigreed canola seed with the Roundup Ready trait. That research indicated that even under the strict segregation protocols used for producing certified seed, contamination was ubiquitous. While his letter to McLachlin could have no effect on the ruling against Schmeiser, Van Acker (2004) expressed his hope that the chief justice would consider his statements "for future rulings and in discussions [she] may have with [her] fellow justices."

As mentioned earlier, I make no claim to know what really happened in Schmeiser's canola fields. Still, in examining this case, the evidence does not look particularly strong that Schmeiser's fields were *intentionally* planted with Roundup Ready canola. The Supreme Court ruling nevertheless effectively closed any scientific controversy, asserting as fact the allegations made by Monsanto. All the Supreme Court justices agreed that Schmeiser in fact had saved and cultivated seeds that he knew to contain the patented genetic trait—although Schmeiser and others continue to challenge the evidence supporting this assertion. This case suggests one way that legal mobilization—the expenditure of social movement resources on legal actions—might backfire. While the courtroom might sometimes be a venue in which science and technology are opened up to public scrutiny, revealing uncertainty and knowledge gaps, it might also be a place where scientific controversies are prematurely put to rest. Here, the Supreme Court, including the dissenting justices, treated the high level of contamination of Schmeiser's field as more solidly factual than the evidence may have warranted. Yet that conclusion mattered only insofar as the Supreme Court believed that patent rights extended to these circumstances, and on that point, the justices were divided.

Defining Genes Out of Place

In the dissenting opinion, four justices argued that a ruling against Schmeiser improperly extended the rights of the patent holders beyond the scope of the patent. They insisted that there is no patent claim "for a 'glyphosate-resistant' [that is, Roundup-resistant] plant and all its offspring. Therefore, saving, planting, or selling seed from glyphosate-resistant plants does not constitute an infringing use" (*Monsanto v. Schmeiser* 2004, par. 162). Indeed, as noted above, Canada does not recognize patents on higher life-forms such as plants. The four judges' colleagues, in the majority, ruled instead that Schmeiser "used" Monsanto's patented gene and thus infringed on the company's patent rights. The disagreement is rooted, in part, in the analogies that the justices chose to apply in making a ruling about the use of the patented invention.

The Supreme Court ruling suggested that merely growing, harvesting, and selling canola seed containing the Roundup Ready gene constituted a use of the invention patented by Monsanto—even though Schmeiser did not take advantage of the plants' resistance to Roundup herbicide in order to control weeds. This assessment was based on the justices' use of several analogies drawn from case law that address the relationship between patented inventions and other objects not explicitly covered by a patent. For instance, as indicated in the ruling, a defendant infringes a patent if he or she "uses a patented part that is contained within something that is not patented, provided the patented part is significant or important." The justices in this case viewed the Roundup Ready gene as important to the structure of the entire plant, in the same way that patented Lego blocks would be important to the structure of an object built with them. They also compared the patented gene to a machine that produces zippers. Under the law, if there is a patent on such a machine, then the patent extends to zippers made by that machine. By analogy, the patented gene produced the herbicide-resistant seeds and plants that Schmeiser used (ibid., par. 41–42).

Four Supreme Court justices strongly disagreed with the analogies used by those in the majority. In a dissenting opinion, they wrote:

There is no genuinely useful analogy between growing a plant in which every cell and every cell of all its progeny are remotely traceable to the genetically modified cell and contain the chimeric gene and putting a zipper in a garment, or tires on a car or constructing with Lego blocks. The analogies are particularly weak when it is considered that the plant can subsequently grow, reproduce, and spread with no further human intervention. (ibid., par. 156)

Brewster Kneen, a prominent critic of the patent system, offered a similar argument against the ruling, pointing out that the use of such analogies revealed "a rather abysmal ignorance of biology." Kneen observed that genes do not manufacture plants in the same sense that a machine manufactures a zipper. Neither is the patented gene necessary for the reproduction of the canola plant. "The majority opinion takes no notice of the self-replicating character of life forms. Nor does it appear to recognize that canola plants have, of necessity, the same growth and reproductive processes whether or not they have been genetically engineered to contain the [Roundup Ready] genetic construct" (Kneen 2004).

These issues were not the only ones concerning the matter of use, though. The ruling also considered something called the "standby" utility of the patented invention. The Supreme Court likened the patented genetic sequence to a patented fire extinguisher that is used even when it is not actually putting out a fire. Here, the Supreme Court compared Schmeiser's possession of the patented gene to a case in which a ship's master possessed patented pumps on board a ship, but had never used them. In that case, the court held that although the pumps had not been used, the ship's master intended to use them if the need arose, and this constituted infringement (*Monsanto v. Schmeiser* 2004, par. 51). Schmeiser could not prove that he never intended to use the Roundup Ready gene in the future; given this, merely the presence of the genes on his land constituted use.

The dissenting judges argued strongly against this conclusion, saying the effect of applying the concept of standby utility in this was to confer patent protection on whole plants—which is not permitted in Canada. They contended that Schmeiser was

entitled to rely on the reasonable expectation that plants, as unpatentable subject matter, fall outside the scope of patent protection. Accordingly, the cultivation of plants containing the patented gene and cell does not constitute an infringement. The plants containing the patented gene can have no stand-by value or utility as my colleagues allege. To conclude otherwise would, in effect, confer patent protection on the plant. (ibid., par. 160)

In sum, the Supreme Court was divided on whether Schmeiser's act of saving, planting, and harvesting canola that contained a patented gene constituted patent infringement under Canadian law. This disagreement was not just a matter of an interpretation of patent law; it also required an interpretation of biology. The justices in the majority believed that analogies to mechanical objects (zipper machines or pumps on ships) were suitable for identifying the reach of Monsanto's patent, while the dissenting justices thought that those analogies created an inaccurate characterization of the relationship between genes and plants.

It is conceivable that had just one more judge sided with the minority view, the outcome of the case would have been different, not only for Schmeiser, but also for the broad public debate about biotechnology in Canada. The dissenting justices made the case that the subject of the lawsuit, especially the matter of "innocent bystanders" to contamination events, deserves broad public debate and "should be expressly considered by Parliament because it can only be inadequately accommodated by the [existing] law on use" (ibid., par. 159). Thus, a ruling in favor of Schmeiser might have prompted a parliamentary debate, inviting broader public engagement in shaping the system of intellectual property rights for plant materials and potentially creating a new legal status for genes out of place. One analyst of the minority opinion makes the following point:

By inquiring into possible responses by Parliament to the patenting of higher life forms or considering alternative conceptions of property, such as that under the PBR, the minority in *Monsanto* produces an alternative discourse of the state in the age of globalization. Canada is cast as an actor with agency which may legitimately develop its own sovereign policy with regard to intellectual property and innovation. (Robertson 2005, 247)

The majority, of course, saw things in an entirely different light. Indeed, rather than imagining an alternative way to govern plant genetic resources, the ruling against Schmeiser used analogies to older, mechanical technologies in order to attribute a new legal status to genes out of place.

Mobilization after *Monsanto v. Schmeiser*

While *Monsanto v. Schmeiser* began as a biotechnology company's effort to exert its intellectual property rights, it became an opportunity for activists to publicize the consequences of releasing GE crops into the environment, and it remains a key point of reference for the antibiotech movement today. While certainly the major impact of *Monsanto v. Schmeiser* was to strengthen the rights of gene patent holders, an additional outcome was that it raised questions about transgene flow to a new level of public scrutiny. For some time prior to the lawsuit, plant scientists had been studying the potential ecological impacts of releasing GE crops into the environment. Advocacy groups concerned about GE crops primarily addressed the topic of transgene flow in terms of the creation of "superweeds"—the hypothesized result of interbreeding between GE crops and wild relatives. Some NGOs, such as Greenpeace, had also raised alarms about the possible contamination of the food supply, through the mixing of GE and non-GE foods. But when Schmeiser claimed that the Roundup Ready canola in his fields must have resulted from pollination and wind-borne seeds, genes out of place took on a new meaning for biotechnology critics. Not only could fields become contaminated, but such contamination could result in patent disputes and severe legal repercussions for farmers.

Rather than closing down activism around plant patents and transgene flow, the ruling in *Monsanto v. Schmeiser*, publicized by the charismatic farmer, his transnational activist allies, and the international media, has ensured that a degree of public controversy continues. Responding to the Schmeiser case and subsequent incidents, some advocacy organizations argued that the biotechnology industry was taking advantage of transgene flow to advance its agenda, waging a strategy of "control by contamination" (DeSantis 2003). Following the ruling, ETC Group called for worldwide activism and initiating a letter-writing campaign. Pat Mooney of ETC Group (2004) commented: "This ruling will unite farmers and others opposed to corporate control of food and life, and galvanize civil society to take the issue out of the courts and back to politicians." And indeed, Schmeiser's struggles contribute to a global debate about GE crops, even as the debate about patenting and transgene flow appears legally and politically closed in Canada.

Schmeiser has become a spokesperson for and "folk hero" in the worldwide opposition to biotechnology. His speaking engagements around the globe have been vital in making the issue of transgenic contamination a central concern for anti-GE activists. Schmeiser is a member of the International Commission on the Future of Food and Agriculture, a project coorganized by world-renowned Indian activist Vandana Shiva. He and his wife have become beloved figures among food and farming activists around the world. In 2007, they were chosen for the Right Livelihood Award, the "alternative Nobel Prize," "for their courage in defending biodiversity and farmers' rights, and challenging the environmental and moral perversity of current interpretations of patent laws" (Right Livelihood Award Foundation 2007).

Environmental organizations and plant genetic resources activists have frequently commented on *Monsanto v. Schmeiser*, and used it to make larger points about the crisis in the agrifood system. For example, Pat Vendetti, a campaigner for Greenpeace in Canada, made these remarks about the Supreme Court ruling: "The decision of the court essentially makes farmers liable to Monsanto for Monsanto's own genetic pollution. It means that Monsanto can reach into farmers' fields and steal their profits and livelihoods" (Greenpeace International 2004). A commentary from Friends of the Earth Europe drew out the following implications:

Percy Schmeiser's case underlines the increasing tension between farmers and large biotech companies, which with their introduction of patented genes intend to change traditional agricultural patterns forever. The impact of these changes on farming communities worldwide could be tremendous. In the South, where people will likely not be able to afford high-tech seeds and the associated chemical inputs year after year, the introduction of GM seed varieties presents a particularly grave threat to the food security and food sovereignty of thousands of local and indigenous farming communities. (Villar 2001)

Since the ruling, Monsanto has asserted that it will not pursue patent infringement cases every time its intellectual property gets loose in the seed supply. Monsanto (2011b) says that "it has never been, nor will it be Monsanto policy to exercise its patent rights where trace amounts of our patented seed or traits are present in farmer's fields as a result of inadvertent means." This does little to appease its critics, however. The plaintiffs in the PUBPAT case, cited in the introduction to this chapter, point out that that this statement is deliberately vague (Public Patent Foundation 2011). How much is a "trace amount"? What exactly are "inadvertent means"? The industry maintains that it is only concerned about *intentional* patent violations, not inadvertent contamination (or "adventitious presence," in industry parlance). But in the Schmeiser case, these lines are not clearly drawn. Should a farmer be restricted from using their own seeds, simply because they become aware that it contains transgenic contamination? In its complaint against Monsanto, PUBPAT also challenges the validity of Monsanto's patents. In contrast to one of the basic requirements of patenting, the plaintiffs, a large coalition of farmers and organizations, contend that transgenic seed is not "useful for society." The PUBPAT case— which at the time of writing this chapter was still in early stages--could establish some limits on the scope of biotechnology patents, creating modest protections for farmers whose crops become contaminated through processes of transgene flow. While the plaintiffs in the PUBPAT lawsuit hope to find some clarity on the matter of patents on genes out of place, it remains to be seen whether the issue can be put to rest without new legislation. The large number of farmers and agricultural businesses involved in the action against Monsanto attests to the widespread anxiety generated by the company's actions surrounding its intellectual property rights. Less clear, however, is whether legal mobilization, using scientific expertise as evidence, is likely to significantly challenge the prevailing social order.

Conclusion

The Schmeiser case offers another example of a scientized debate about biotechnology. Like the Mexican maize activists, Schmeiser came to depend heavily on genetic testing for transgenic material in defense of his seed saving practices. And in both cases, while farmers and activists sought to remake agricultural policy—here, to challenge the system of intellectual property rights for plant breeders—formal decision-making processes left those larger questions off the table. When Monsanto and Schmeiser met in the courtroom, the contest hinged on scientific evidence that Schmeiser's fields contained so many plants with the Roundup Ready gene that the seeds could not have gotten there accidentally. Scientific questions took center stage in a case where experts squared off against counterexperts about the quality of the data collected from Schmeiser's canola fields. After the ruling, some critics questioned the Supreme Court justices' grasp of the science pertinent to the case. But legal proceedings do not allow the same kind of uncertainty and correction process that is more typical of the scientific field. The court's ruling effectively closed the scientific controversy, while foreclosing the possibility of further legislative debate.

The case offers some lessons about science and the law as part of contentious politics. Science can be a valuable ally to activists in the courtroom. In legal proceedings, critics can use scientific counterexperts to challenge claims about the safety of a new technology. However, science is an "unreliable" ally to social movements because it moves slowly and is subject to revision with new evidence (Yearley 1992). In a legal conflict, science is particularly undependable because it is difficult to predict how judges or a jury will interpret uncertain or contradictory scientific knowledge.

Clearly, science is not a "magic wand" that resolves complex legal disputes (Caudill and LaRue 2006). Nevertheless, mobilization in the courts—both as defendants and plaintiffs—is likely to remain a central strategy for the antibiotech movement. In a context in which regulatory agencies will consider only a narrow range of scientific questions and the legislature is reluctant to challenge a powerful industry, the judicial system is an alternative venue in which challengers can seek to address the ambiguous legal status of genes out of place. The impacts of legal mobilization, though, are more likely to stem from the broader campaign surrounding a court case than from a judge's ruling. *Monsanto v. Schmeiser*—initially just one of many cases in which Monsanto made patent infringement allegations against farmers—became a venue for strenuously voicing objections to "life patents" and defending farmers' rights to save seeds. The ruling was clearly a loss for the antibiotech movement, but in the context of broader social resistance to the dominance of Monsanto and other multinational agribusiness corporations, it fueled worldwide debate and subsequent legal challenges.

Schmeiser's conflict with Monsanto highlighted what some see as a contradiction in Canadian law. As legal scholar Martin Phillipson (2001) points out, even while granting patent holders extensive rights, the Canadian courts have not indicated that owners of such patents have any corresponding obligations (see also de Beer 2007). For instance, Monsanto may claim ownership of a gene present because of cross-pollination in an organic farmer's field, but the company appears to be under no legal obligation to prevent its genes from getting where they are not wanted. There have been some legal actions to try to address this apparent imbalance.

After the Supreme Court ruling, Schmeiser and his wife, Louise, themselves pursued—and won—a small claims lawsuit against Monsanto, accusing the company of contaminating their farm with GE canola. The case was settled out of court in 2008, with Monsanto agreeing to pay for the removal of Roundup Ready canola from the Schmeisers' land (Provincial Court of Saskatchewan 2005; Hartley 2008). Organic farmers in Saskatchewan also confronted Monsanto in court, accusing the company of destroying the prospects for producing organic canola. That legal battle, in which farmers asked the courts to treat genes out of place as analogous to pollution, trespassing, or a nuisance, is the subject of the next chapter.

6

Protecting Organic Markets

The public rejection of GE foods, in places as varied as the European Union, Brazil, and Japan, has limited the global adoption of GE seeds, while simultaneously creating new opportunities to market non-GE foods.[1] In some countries, labels are required on foods containing GE material. Additionally, certified organic products, grown without GE seeds or other inputs, have captured a growing share of the global food market.[2] Farmers who produce non-GE crops are vulnerable, though, given the difficulty of avoiding unwanted contamination with GE material. In Saskatchewan, Canada, while most farmers were adopting herbicide-resistant canola, some organic farmers took advantage of the international demand for non-GE canola, receiving a premium for certified organic crops. But when transgene flow made it nearly impossible to produce a crop of canola that did not contain GE material, buyers no longer wanted what the Saskatchewan organic farmers had to sell.

A committee called the Organic Agriculture Protection Fund (OAPF) organized a class action lawsuit against Bayer and Monsanto, the makers of transgenes that were contaminating organic famers' fields. The organic farmers sought to apply existing law to genes out of place, treating them as analogous to other objects that are governed by the state, such as environmental pollution or trespassers. The lawsuit highlighted contradictions between the patent system, which treats transgenes as patented inventions, and the regulatory system, which views them as a benign addition to the environment. The plaintiffs asked whether biotechnology companies, having been granted expansive patent rights in *Monsanto v. Schmeiser*, have any corresponding responsibilities to monitor and control those inventions. In its response to the organic farmers' claims, Monsanto argued that "there is no way to put the genie back into the bottle. GM canola is now part of the environment. Our federal government authorized and approved that" (Garforth and Ainslie 2006, 473). The organic farmers pointed out the contradictions in Monsanto's position, asking, Are they inventions, or are they, as Monsanto maintained, "part of the environment"? What legal categories apply to these objects?[3]

In this chapter, I examine the OAPF lawsuit and the political mobilization surrounding it. Multi-institutional challenges to GE crops, not only in Canada, but worldwide, are shifting the "rules of the game" governing the development of biotechnology. The Canadian government has offered few protections for organic or other non-GE producers, indicating that if farmers wish to avoid GE crops once the government has deemed them safe to commercialize, it is their own private responsibility to do so. It is in this context that organic farmers have found themselves in direct confrontation with the biotechnology industry and the government. The organic farmers in this case assert that government deregulation of GE crops poses an economic threat to them, by diminishing the marketability of their crops.

While the biotechnology industry continually advocates "science-based" decisions, farmer organizations and antibiotech activists in Canada have called on the government to include an assessment of the marketability of GE crops in regulatory decisionmaking. I show how the institutions of the organic industry—while seemingly aligned with neoliberal ideas of consumer choice rather than government intervention—are a resource for struggles to change the dominant institutions of agricultural governance. Specifically, organic standards that forbid GE seeds—and the extension of these standards, in practice, to require the production of non-GE foods—has become the basis for demands that the government rein in the power of the biotechnology industry. These efforts are strongly influenced by antibiotech activism in Europe and other export markets, where consumer activists have pressured supermarkets to adopt prohibitions on GE foods and governments have created strict product labeling rules. In turn, Canadian farmers and activists have targeted multiple institutions, including various branches of the government and the biotechnology industry, in their demands to remove GE seeds from the market. The uneven success of these efforts demonstrates the durability and widely taken-for-granted nature of the idea that governance of GE crops must be based only on scientific risk assessment, despite clear failures of the regulatory system.

I begin with a discussion of several efforts to introduce a marketability criterion for GE crop regulation in Canada, followed by an explanation of the specific problems that genes out of place pose for organic farmers. This is followed by an analysis of the outcome of the OAPF class action, in which I consider how the legal system constrains the possibilities for conceptualizing genes out of place and their consequences. Finally, I consider where the marketability claim—used by organic farmers to challenge the scientization of biotechnology governance and advance a more radical argument against GE crops in all agriculture—might lead.

Consumer Activism and Marketability

Canadian farmers who depend on export markets must be responsive not only to foreign government regulation of GE foods but also the policies of companies that process and sell food. In Europe, for example, consumer activism in the 1990s led many food processors and retailers to reject all GE foods. In some cases, opposition to GE foods in key export markets has led broad coalitions of farmers to reject GE seeds, forcing seed producers to withdraw those seeds from the market. But when Canadian legislators and regulatory agencies have considered the possibility of formalizing an assessment of the marketability of GE crops, they have rejected the idea in favor of a science-based approach to regulation. For both farmers and food consumers, therefore, nonstate institutions (seed and food companies) are important targets for antibiotech challenges. It is increasingly evident, though, that market-based forms of governance of GE crops (such as voluntary seed withdrawals and organic standards) are ineffective at resolving conflicts over genes out of place.

Since the 1990s, widespread public opposition and a precautionary approach to regulating GE foods in Europe have severely limited imports of agricultural products from the United States, Canada, and other GE crop-producing countries. Responding to strong consumer-centered campaigns, from March 1998 to June 1999, at least fourteen of the largest supermarket chains and food processors in Europe announced that they would no longer sell foods made with GE ingredients (Schurman 2004, 256). The antibiotech campaigns also spurred governments to action. In June 1999, the European Union as a whole changed its regulatory framework, moving toward a precautionary approach. Given the already-high distrust of public authorities as a result of several major food scares throughout the 1990s, western European governments came to see that their legitimacy rested on responding to consumer objections to GE foods (ibid., 262). In 2003, the European Union put into place a system for ensuring the traceability and labeling of GE crops throughout the food supply chain (Lezaun 2006). A 2005 study by Greenpeace discovered that even prior to the institution of those requirements, European food processors and retailers had rejected nearly all GE foods and ingredients (Holbach and Keenan 2005). The same strategy of targeting the food processors and retailers has also been successful in other parts of the world, such as Russia, where two major food and feed importers announced in 2006 that they would use only non-GE products (Greenpeace International 2006).

Currently, Canadian farmers bear the full responsibility of ensuring that their crops remain free of GE material, if they wish to do so. As described in chapter 2, Canada's research and regulatory systems for GE crops have made the country one of the major producers of GE crops. Canadian law makes no explicit statements about the obligation to protect farmers' rights to grow non-GE crops. Indeed, under the conventions of the Canadian legal system, growers who wish to capture a price premium for non-GE crops must themselves bear the costs and responsibilities of meeting the requirements of customers and certification bodies (Smyth and Kershen 2006, 8). This system contrasts with the "coexistence" systems that are being developed in Europe, which aim to allow the pursuit of both GE and non-GE (including organic) agriculture within a country. The Commission of the European Communities (2003) introduced coexistence guidelines in 2003, and several European countries have used these guidelines to develop their own coexistence policies. Because cross-pollination is considered inevitable, these policies set a threshold for acceptable levels of contamination and provide guidelines for keeping contamination below that threshold (such as field-spacing distances).

Canada's position exhibits greater similarity to the policies of the United States. The United States, like Canada, has declined to take a role in protecting non-GE crop production. The recent decision, discussed in this book's introduction, to forego coexistence provisions for GE alfalfa, advanced the US position that protection of non-GE crop production is not a government responsibility. In the United States, liability for contamination has been raised in a number of lawsuits, but the courts have been disinclined to consider liability for transgene flow.[4] The US courts have ruled that the only situation in which liability is relevant to biotechnology is in cases such as the StarLink scandal, in which unapproved GE corn accidentally entered the food supply.

When StarLink became mixed with corn for human food, both farmers and grain elevators filed lawsuits against the developer of the GE corn, Aventis Cropscience USA. These were joined into a single class action lawsuit. The trial judge ruled that "plaintiffs who could prove that their crop or stored grain had been physically contaminated by unapproved StarLink—making their crops and grain unmarketable as food corn because *adulterated by an unapproved substance*—had a viable legal claim through negligence, private nuisance, and public nuisance" (Smyth and Kershen 2006, 6–7; emphasis added). In other cases, however, such as a class action brought by growers of conventional soybeans and corn against transgenic seed developers (*Sample v. Monsanto Co.*), the US courts determined that biotech companies were only liable for direct physical injuries to crops, not loss of markets due to contamination of non-GE crops (ibid.,13–14). This appears to be the position of the US courts even if the crops are unapproved in other countries. Individual states have also opposed the creation of protections for non-GE agriculture. At least thirteen state governments have passed legislation prohibiting localities from creating "GM-free zones" (Endres and Redick 2006, 3).

In Canada, as well as the United States, where there are no formal coexistence rules, decisions to commercialize GE seeds affect all producers, not just those who voluntarily grow GE crops. Companies like Monsanto are clearly aware of this, and have been strategic in introducing GE crops in ways that will gain the greatest market share and avoid massive consumer rejection. When the biotechnology industry first introduced GE canola in Canada, the companies implemented identity preservation systems (Phillips and Smyth 2004). That is, separate growing, transportation, and processing operations were created for GE and non-GE canola, until it was certain that there was a large enough market for GE varieties around the world (besides Europe). The identity-preservation practices were then abandoned.

There are many examples in which the rejection of GE foods in export markets led to withdrawals of GE seeds from the seed supply in Canada. In one case, GE flax, developed by researchers at the University of Saskatchewan, was removed from the market and deregistered after flax farmers' and producers' groups insisted that it threatened their ability to sell flax in Europe. A representative of the Canadian Food Inspection Agency (CFIA) explained, "There was nothing wrong with the variety. It met all the requirements, [but] they all agreed the variety should be deregistered. This is unique" (Warick 2001). Pressure from farmers, concerned about losing customers, prompted the government to deregister the seed variety in this instance (Eaton 2011, 124–125). In fact, from 1990 to 2002, the CFIA had the power to evaluate the economic impact of a new crop, but few people, including employees of the agency, knew of the clause that empowered CFIA in this way. A report based on an Access to Information Request revealed that when agency employees discovered the clause in 2002, CFIA officials removed it, giving up this power (Warick 2003).[5]

In subsequent cases, the government has had a more restricted role, allowing the biotechnology industry to handle matters of marketability on a voluntary basis. Monsanto announced plans in 2001 to introduce GE herbicide-resistant wheat in Canada. When important wheat millers in Japan and the United Kingdom said that they would not purchase wheat with any level of GE contamination, many Canadian farmers turned against the idea (Olson 2005, 157). Some opponents of GE wheat articulated a multidimensional critique, saying that biotechnology companies have an inordinate amount of power, the regulatory system is fundamentally flawed, and there is a need for greater transparency and democracy (Magnan 2007; Eaton 2009). Focusing on marketability, though, allowed coalition building across diverse organizations and interests. Throughout the mobilization against Roundup Ready wheat, as in the earlier cases of flax and canola, the dominant frame or way of defining the problem was potential harm to farmers through "market loss," based on the idea that the "consumer knows best" (Eaton 2009, 272; Magnan 2007, 306).

In a remarkable example of coalition building, environmental groups, consumer groups, and farm-sector organizations from across the political spectrum joined forces to demand that Monsanto withdraw its application to market GE wheat in Canada.[6] Responding to this pressure, in January 2004, Agriculture and Agri-Food Canada (AAFC), a public institution that for years had been collaborating with Monsanto to develop and test Roundup Ready canola, withdrew from the project, citing concerns about the marketability of the transgenic wheat. Then, under continued pressure, Monsanto (2004) announced in May that after consultation with growers, it was deferring field research and breeding of Roundup Ready wheat (Monsanto Company 2004). It seems likely that Monsanto will attempt to introduce transgenic wheat again in the future; nevertheless, critics of transgenic crops cautiously welcomed the company's withdrawal as a victory.[7]

Rejection by global markets because of crop contamination became a reality in 2009, in a controversy involving Canadian-grown flax. That year, there was a series of puzzling discoveries of food containing the very type of GE flax that had been officially deregistered and destroyed in 2001, but appears to have persisted in the flax seed supply. The contamination was discovered in genetic tests of foods shipped to Europe and Brazil throughout 2009, and importing countries halted trade in Canadian flax (Greenpeace International and GeneWatch UK 2009; Nickel 2010). The controversy over contaminated flax propelled a legislative proposal to require market analysis before commercializing GE seeds. In 2010, Alex Atamanenko, a member of Parliament from British Columbia, introduced Bill C-474, which would amend the Seeds Regulations to require an analysis of potential harm to export markets before the sale of any new GE seed is permitted. Organic farmers were among the supporters of this regulation. A writer on behalf of Canadian Organic Growers, an organization that advocates for organic farmers in Canada, voiced support for the legislation:

Currently, our national regulatory system is deeply flawed, and is arguably designed to benefit corporations that develop GE crops at the expense of organic farmers and consumers as a whole. Market impact must be included in the overall assessment of this technology. Indeed, history has shown us the dangers associated with a regulatory system that is solely "science-based," and we now know that the current system is too narrow to properly evaluate the multitude of potentially adverse socio-economic impacts associated with this technology. (Taylor 2010)

The letter indicated that methods for assessing market impacts could be based on "cost-benefit analysis tools" that were developed by the Canadian Wheat Board to evaluate the effects of GE wheat and on the "farmer-focused risk assessment" methods used by environmental researcher Ian Mauro (Mauro and McLachlan 2008; Mauro, McLachlan, and Van Acker 2009), who with his colleagues showed that farmers perceive marketability as a crucial factor in assessing the risks of GE crops.

The proposal was strongly opposed by the biotechnology industry and agricultural industry groups that have been supportive of GE crops. For example, the Canadian Canola Growers Association argued that the bill would discourage biotechnology companies from investing in new research because they would be uncertain about whether seeds will be approved under the new framework. That is, market assessments were viewed as less predictable than the existing regulatory framework. As seen in episodes involving GE maize and other transgenic crops, industry objections to government intervention were again framed as a desire to maintain science-based regulation. A statement from the industry lobby group CropLife Canada (2010; emphasis added) said that proposal

is bad news for farmers, consumers and Canada's agricultural exports because, if passed, the Bill would put *highly subjective and non-scientific criteria* into the regulations governing innovation [in] plant biotechnologies. Moving away from science-based regulation leaves our exports vulnerable to frivolous trade challenges and limits our ability to adopt valuable innovations for all Canadians.

Lucy Sharratt, a prominent Canadian critic of GE crops, countered that market assessment would be "science-based" and "fact-driven," as it would involve a systematic review of the status of GE food imports in the countries that Canada serves (Shane 2010). This argument, however, did not seem to convince legislators.

The Canadian government has frequently echoed the industry's call for science-based regulation, indicated in chapter 2. For instance, in the negotiations for the Cartagena Biosafety Protocol throughout the 1990s, representatives from the Global South called for the right to regulate GE crops based on the expected socioeconomic impacts. In those negotiations, as in other conflicts over trade in biotechnology, Canada argued firmly that regulations must be based only on scientific assessments, not market considerations (Kleinman and Kinchy 2007; Andrée 2007). Canada has also used the WTO to challenge the European Union for creating barriers to trade in GE crops that were not sufficiently based in science. It is therefore not surprising that the majority of legislators supported the industry's view and ultimately voted down Bill C-474. One member of Parliament explained, "Our government, along with the vast majority of farmers and industry leaders, supports a safety approval process based solely on sound science," insisting that the existing regulatory process protects consumers and the environment (Cockrall-King 2011). Thus, while companies like Monsanto might voluntarily halt the commercialization of GE seeds under pressure from a wide swath of producers, legislators have largely opposed an increase in government oversight.

In this context, organic farmers—lacking support from conventional farmers, who largely favored GE canola and had a market for it—found themselves without allies and unprotected by the government. As one advocate of organic agriculture put it, "People had seen how the regulatory system was not protecting us, and the political system wasn't protecting us, so it seemed that the only hope we had was the courts" (food and farming activist, Saskatoon, July 13, 2006). Organic farmers would find, however, that the legal system was also disinclined to recognize their objections to GE crops. Before presenting the details of the lawsuit, it is necessary to consider what was at stake for the organic farmers who chose to take legal action.

Organic Farming and Genes out of Place

Organic farming is perhaps the best-known form of alternative agriculture, and organic farmers have recently been at the forefront of battles against GE crops. Because organic farming has formal rules, standards, product labels, certification organizations, processing companies, and brands, it offers a productive site to examine how critics of agricultural biotechnology might use these alternative institutions to press for change in the dominant institutions of agricultural governance. Some analyses have concluded that organic agriculture has lost its transformative potential. Julie Guthman (2004, 2007), a geographer who studies the politics of food and agriculture, argues convincingly that the codification of organic standards and development of international markets for organic commodities have followed neoliberal principles of governance, relying on consumer choice to address the ecological harms caused by industrial agriculture. As a consequence, organic has become less a movement to transform industrial agriculture and more an elite market niche. Indeed, organic farmers frequently describe GE crops as economically harmful, because they threaten the marketability of the commodities they produce. The dominance of marketability as a way for organic farmers to express their grievances reflects their sensitivity to consumer demands for non-GE products. It also seems to suggest that organic farming is an industry looking out for its own survival, not a movement for social change.

Yet the distinction between these two impulses—defending markets and advocating for social change—is often blurred. Writing from a US perspective, critical scientist Martha Herbert (2005, 66) observes:

The assertion of this right [to food that is not GE] is much more than a meek demand for little preserves or reservations of organic farming in the midst of vast spreads of GE crops, or a tame request for GE-free labels on our food and GE-free aisles in our supermarkets. Certainly, demands for protecting organic farming and for food labeling have tactical importance. But they are not enough.

The institutions of organic agriculture, such as labels and standards, only offer limited potential for transformative social change. The case of the Saskatchewan organic farmers reveals that the difficulty of containing GE crops once they are released into the environment has surpassed the capacity of voluntary, market-based systems. In this context, Canadian organic farmers and antibiotech activists have fought hard for the government to treat genes out of place as it would pollutants, nuisances, or other harmful substances. The defense of organic agriculture thus has become explicitly political.

Over the past several decades, organic agriculture has become the most visible alternative to industrial farming systems in the Global North, and organic production has grown rapidly in Canada. In 2008, there were over a thousand certified organic farms in Saskatchewan, representing 2.3 percent of farms in the province (Agriculture and Agri-Food Canada 2010). Although organic farming is often associated with fruit and vegetable production, more than half of the certified organic farms in Canada produce field crops such as wheat and oats. Wheat is the primary organic crop produced in Saskatchewan (ibid.). Many organic farmers in the province are members of Saskatchewan Organic Directorate (SOD), a nonprofit organization that promotes organic agriculture through research, education, advocacy, and information sharing. SOD helps farmers transition to organic farming by identifying certification agencies, facilitating the exchange of farming skills, and advocating for the benefits of organic production. Interviewing organic farmers in Saskatchewan, I found that environmental and health concerns were not typically the sole or primary reason for adopting organic practices. One organic farmer (Saskatchewan, July 14, 2006), for example, said that in addition to not being happy about spraying chemicals, "I had never been very happy about the sort of corporate control that was happening to farming. And the reality is, if you're into high-input agriculture, you're for the likes of Monsanto, so that really rubbed me the wrong way. So it was a way of trying to get out from underneath the thumb of sending your profits off to a middleman."

Organic farming got its start in Saskatchewan in 1969, when Elmer and Gladys Laird transitioned their farm to organic methods in order to cut the high cost of chemical inputs.[8] Organic farming was gaining interest in Canada at the time.[9] In the 1970s, organic organizations were started in several provinces, and a small number of university researchers began studying and teaching about environmentally sustainable agriculture. At the same time, organic agriculture was rapidly developing internationally, with the International Federation of Organic Agriculture Movements, established in 1972, helping to link organic advocates around the world.[10] The organic industry began to take its contemporary shape in the 1980s, as organic organizations started to create certifying agencies, standards, and product labels. The Organic Trade Association was formed in 1985 to represent the interests of organic businesses. By the end of the 1980s, the Canadian government had begun to take an interest in organic agriculture, funding research programs on sustainable agriculture and initiating discussions about the regulation of the label "organic."[11]

As organic standards have been formalized and brought under the regulation of national governments, they have reduced the organic principles to a set of rules regarding permitted and forbidden substances to be used on the farm (Guthman 2004). In particular, organic standards forbid the use of synthetic chemicals, fertilizers, or GE organisms. Organic advocates have long opposed GE crops, arguing that they pose an ecological threat to the planet, threaten human health, and endanger the economic independence of farmers. The International Federation of Organic Agriculture Movements (2002), for example, advocates a total, worldwide ban on genetic engineering.

Keeping GE organisms out of organic food was a major campaign in the United States in the late 1990s, when the USDA proposed national organic standards that would permit the use of GE seeds. Responding to a large public outcry against the proposal, the USDA ultimately created national organic standards that forbid the use of GE seeds. Yet some critics of biotechnology and industrial agriculture looked skeptically at the extent of this victory. For example, David Goodman (2000, 213), a prominent scholar of agrifood politics, observed that with the new organic standards, "no progress has been made towards addressing the social objectives of sustainable agriculture and food systems, whether the social relations of the farm labor process or equitable access to safe nutritious food." Instead, the new organic standards are "entirely consistent with neo-liberal approaches to industrial regulation," leaving decisions about the desirability of biotechnology up to "informed consumers" and the market (ibid). Canadian organic standards closely resemble those in the United States, and the same critiques apply. The Saskatchewan organic farmers' case, however, indicates that consumer demand can serve as leverage for demands for government regulation.

All organic standards are based on process (how a crop was grown), not product (what is in or on the food), so even if some accidental contamination does occur due to windblown pollen or accidentally scattered seeds, it should—according to the standards—still be considered organic. In the case of pesticides, for example, the organic label is not meant to guarantee that there has been no pesticide drift from neighboring fields, although consumers might assume that organic produce is completely free of pesticide residues. A similar situation holds with respect to transgenic material. Organic standards in Canada require farmers not to use GE seeds. Although this standard does not guarantee that a product will be completely non-GE, genetic testing is now widely used, particularly in Europe, where traceability systems have been in place since 2003. A positive test for contamination can raise doubts in a buyer's mind about a farmer's practices. It can also mean rejection by European food processors and retailers that have pledged to sell only non-GE food.

Given these pressures, organic producers have become attuned not only to adhering to an organic process but also to producing a non-GE product. One Saskatchewan farmer, Dale Beaudoin (Saskatchewan, July

20, 2006), described his experience with this expectation. In the 1990s, he grew organic canola, typically under a contract with a buyer from the United States. He heard about a buyer in 1999 in Quebec that was offering contracts for organic canola at a much better price: $16.50 a bushel. But when the crop was ready to sell, the buyer tested a sample and found that it contained some transgenic seeds, and therefore did not want it. Beaudoin ended up finding another buyer for his canola crop, but at a much lower price: $11 per bushel.

Faced with the threat of losing their ability to produce non-GE products, organic farmers have sought protection from the Canadian government in a variety of ways, including taking legal action. By the end of 1999, a number of organic farmers in Saskatchewan were talking about the problem of contaminated canola, and some decided they could no longer grow it. Then Monsanto, the maker of the most widely grown transgenic canola in Canada, as noted earlier, made moves to introduce Roundup Ready wheat (Olson 2005). SOD followed these developments closely. Organic farmers objected to the risks associated with genetically engineering a staple food crop and feared that the authorization of GE wheat in Canada would severely damage their ability to sell organic wheat on the global market. A group of people from SOD attempted to lobby the provincial government, meeting with the minister of agriculture and threatening legal action if something was not done to protect organic wheat from GE contamination. In 2001, Arnold Taylor, an organic farmer who had given up canola production, became SOD's president and started looking for ways to back up that threat of legal action.

SOD formed a committee—the OAPF—to initiate a class action lawsuit against the two biotechnology corporations that produce transgenic canola grown in western Canada: Monsanto Canada, Inc., and Bayer CropScience, Inc. The committee sought out other farmers who had suffered losses because of the canola contamination. Larry Hoffman, a farmer who had stopped growing canola after hearing about contamination cases, became one of the plaintiffs. Like Taylor, Hoffman described dropping canola from his crop rotation as a disappointment. In interviews, both farmers emphasized that organic canola had been a profitable crop, fairly easy to grow and harvest. For these farmers, canola contamination constituted a loss, both economically and in the sense of losing an option that fit well into their organic crop rotation.[12] Beaudoin also agreed to be one of the plaintiffs. The lawsuit sought damages from the companies whose patented transgenes are now widespread in canola seeds and plants grown across the Canadian prairies. They also sought an injunction against the introduction of transgenic wheat. But in 2004, when Monsanto voluntarily put Roundup Ready wheat on hold, that portion of the lawsuit was dropped.

The farmers involved in the OAPF class action did not necessarily see themselves as activists, reserving that label for those who take more radical actions such as destroying fields of GE crops. As Taylor (Saskatchewan, July 18, 2006) explained, "For some people it's political, but for me it's strictly, almost strictly business, where we're protecting our position in the marketplace. We have a crop [canola] that was taken away—stolen away from right under our noses." For others, the "business" perspective included anger at the biotechnology companies that were gaining increasing power and control over agricultural production in Canada, thereby threatening the organic business. Hoffman (Saskatchewan, July 25, 2006), for instance, said he was not an activist but instead that part of the motivation to take the companies to court was that he and others felt the biotechnology industry had too much power:

We were saying . . . these companies seem to be getting away with just about anything they want to. What do you think about doing something . . . to get some equal rights here? Because it seems like they want their way. They want the cake and to eat it too, and they want to stomp on organic farmers.

For these farmers, defending their business meant not just operating within a market niche but also seeking to change, or at least limit, the practices of much more powerful corporate actors. Therefore, while their grievances were typically expressed in terms consistent with neoliberal ideas of consumer choice and marketability, their demands were somewhat more radical: calling for state intervention to rein in a powerful industry.

One of places where it is evident that Saskatchewan organic farmers directly challenged the dominant policies and practices of Canadian agriculture is in the disagreement about the "tolerance" of contamination or adventitious presence of GE material. I interviewed representatives of the biotechnology industry, biotechnology researchers, and government officials who viewed the organic industry's position as unrealistic. This quotation, from an agricultural scientist working for the Canadian government, conveys a point of view I also heard from others who saw GE canola as a beneficial contribution to Canadian agriculture:

Two years after the first release of biotech canola [organic farmers decided] that they would not accept any. In other words, it was zero transgenic material in organic foods. This was not sanctioned by any government at that time or anybody but themselves. Who are they to decide that that's the level? And is that level realistic? Is it attainable? I think the answer to that is no.

He went on to articulate a position that advocates of biotechnology frequently take: if organic farmers wish to continue to sell a non-GE product, it is their responsibility to find a way to do so, without imposing any restrictions on other farmers. As he put it:

You also have to look at the farmer's liability, and what sort of operation is he running. Was he in part responsible [for contamination]? Did he talk to his neighbors? Are his rights greater than his neighbor as to what crop he is going to grow? Or, you know, is the organic grower going to limit the profitability of his neighbor by saying you can't grow canola anywhere near my farm? (government scientist, Saskatoon, July 13, 2006)

In line with this perspective, advocates of GE crops indicated that consumers should accept a certain level of adventitious presence of GE material. They maintained that objections to adventitious presence were irrational, given that Canadian authorities had determined GE crops to be safe.

In contrast, organic farmers felt that their rights to farm as they wish on their own land were being compromised, and that this right ought to trump the biotechnology industry's right to commercialize GE seeds. Organic farmers I interviewed in Saskatchewan contended that setting a tolerance level for adventitious presence would be a "slippery slope" toward greater acceptance of GE crops in organic agriculture—an outcome they sought to avoid. Those involved in the OAPF class action indicated that GE food was unsafe to eat, harmful to the environment, and not useful for organic food production. Many also objected to the growing power of the biotechnology industry and saw their struggle as more expansive than just a debate about accepting GE content in organic food. As one active member of OAPF articulated: "The GMO issue [is] just a symptom of the corporate takeover of agriculture by various means." He went on to say that he believed accepting a certain level of contamination "will undermine the credibility of the organic farming," and continued: "Now, if we get to a point where contamination is rampant, then what other choices are we going to have? We'll maybe have to accept a level of contamination in organically produced food. But until that day comes, I think the organic sector should really draw a line in the sand and say no way, we're not going there" (organic farmer, Saskatchewan, July 14, 2006).[13]

Genes Out of Place on Trial

One analyst of the farmers' class action describes it as "an extremely blunt instrument" for trying to defend "a multitude of social, economic and environmental values" (Garforth and Ainslie 2006, 476). On the surface, the case focuses on the loss of organic sales due to contamination, but it is clear—in both interviews and legal documents—that they were questioning, at a deeper level, both the power of the biotech industry and the adequacy of the regulatory system. It is tempting to try to distinguish the safety from the socioeconomic issues that are at stake in this instance, yet they are inseparable, not only in the farmers' perceptions of the value of organic farming, but also in the way that the courts assessed whether the case should move forward. In particular, because the judge assumed that GE crops were safe, she seemed to view the farmers' economic grievances as illegitimate. The judge, by the same token, was disinclined to view the farmers as an identifiable class with respect to environmental law, arguably because their claims about economic losses highlighted their individual (not collective) interests. I will elaborate on these points further, after a short explanation of the claims that the farmers made.

I have already discussed the plaintiffs' desire to rein in the power of the biotechnology industry. In combination with this demand, a critique of the safety of GE canola was implicit in the plaintiffs' assertions about the loss of organic markets. As others have pointed out, "lingering questions over the potential for GM crops and food to cause harm to humans and the environment" are at the heart of why "organic certification standards have rejected GMOs in the first place" (Garforth and Ainslie 2006, 468). Despite the CFIA's approval of GE canola, many farmers and consumers worldwide remain unconvinced that it is safe for the environment or healthy to consume. As discussed in chapter 2, the Royal Society of Canada and other official bodies have criticized Canada's regulatory system for GE crops, leaving substantial room for skepticism about the government's risk assessments. Thus, when organic farmers contended that GE contamination damaged their capacity to produce organic canola, they were suggesting not merely that some consumers prefer non-GE canola oil but rather a concern that GE seeds can do physical harm. As part of their claim, the plaintiffs questioned the safety of GE canola, directly challenging the Canadian government's risk assessment determination.

The legal action began in January 2002, when Saskatchewan organic farmers registered a statement of claim against Monsanto and Aventis (now Bayer CropScience). Later that year, Hoffman and Beaudoin filed a motion for certification, asking the judge to recognize them as representing a class, which they defined as all organic grain farmers in Saskatchewan who were certified organic between 1996 and the time of the class certification. All members of this class, they argued, were negatively affected by the introduction of GE canola into the environment. In its statement of claim, the OAPF maintained that Monsanto and Bayer were liable in multiple ways, including nuisance, negligence, trespass, strict liability, and breach of environmental protection statutes.

The organic farmers had to express their grievances with the biotechnology industry using available legal concepts. It was then up to the judge to determine whether any of their allegations constituted a reasonable cause of action—a reason to go to trial. First, the plaintiff farmers contended that Monsanto and Bayer violated provincial environmental law, and hence were liable for the resultant damages. In an article considering the legal ramifications of "gene wandering" in Canada, legal scholar Jane Matthews Glenn (2004) has argued that non-GE farmers who have experienced GE contamination might be able to claim damages for the breach of duties under provincial environmental protection legislation. Under Saskatchewan's Environmental Management and Protection Act, a person is entitled to compensation from someone who causes loss or damage through the discharge of a "pollutant" or "substance" into the environment. Legal scholar Martin Phillipson (2001) has asserted that this is the most promising avenue for organic farmers to pursue the issue of transgenic contamination. The definition of pollutant appears to be broad enough to include transgenes, but there is a legal question as to whether the biotechnology company can be considered the "owner or person in control" of the pollutant, as described under the law. In light of *Monsanto v. Schmeiser*, however, a case could be made that the owners of patents on transgenes are responsible for the pollutant. Indeed, this is the argument that was made by the plaintiffs in this case.

In addition to the Environmental Management and Protection Act, Saskatchewan's Environmental Assessment Act is relevant to the issue of GE contamination. Under this legislation, someone may be held liable to a person who suffers loss, damage, or injury as a result of a "development" for which an environmental impact assessment should have been conducted (Glenn 2004, 559). Glenn convincingly maintains that GE crops should count as a development under the multiple criteria set out in the Environmental Assessment Act. Therefore, in *Hoffman v. Monsanto*, "the argument is that the testing and subsequent unconfined commercial release of genetically modified organisms (such as canola and wheat) is a development within the meaning of the Act for which provincial ministerial approval is required" (ibid., 559–560). Saskatchewan's minister of environment and resource management has not given approval for the dissemination of GE seeds in the province, and non-GE farmers, both conventional and organic, have suffered damages as a result of crop contamination. The plaintiffs in *Hoffman v. Monsanto* held that under the act, Monsanto and Bayer were liable for those damages.

The farmers attempted to apply the law on trespassing to genes out of place, saying that the biotechnology companies were responsible for the unauthorized, but foreseeable entry of those genes on to the organic farmers' land (Garforth and Ainslie 2006). Furthermore, the plaintiff farmers accused Monsanto and Bayer of nuisance. Nuisance involves the unreasonable interference with a person's "use and enjoyment" of their land. Nuisance law typically would be used to recover damages for harm caused by a neighboring landowner. But as Phillipson (2001) explains, in Canada, "right to farm" legislation prohibits the use of nuisance actions against farmers who are using "normally accepted agricultural practices," such as GE crops. In *Hoffman v. Monsanto*, though, the plaintiffs used nuisance law against the manufacturers of transgenes, the biotechnology companies, rather than individual farmers.[14]

In addition, the farmers indicated that Monsanto and Bayer were negligent "under a well-accepted 'products liability' line of argument" (Glenn 2004, 561). There are two dimensions of this line of reasoning that are relevant to GE crops. The first involves a "design defect," meaning that the design of the product itself creates an excessive risk of injury. Legal precedents involving this type of negligence include "thalidomide pills, silicone gel breast implants, even cigarettes" (ibid., 563). Glenn (ibid.) believes that this is the most relevant category of negligence for GE crops, but that it is the most difficult to prove, which may explain why the organic farmers focused on another form of negligence—the concept of "labeling defect." Here, a manufacturer may be liable for damages if it fails to warn of the hidden dangers inherent in reasonably foreseeable uses of the product. In *Hoffman v. Monsanto*, the plaintiffs held that there was a labeling defect in the marketing of GE canola, because farmers were not told to take precautions to prevent the spread of GE pollen and seeds to unintended locations. More specifically, they alleged that Monsanto and Bayer had the duty to warn growers to limit the spread of the gene as well as a duty to maintain the identity-preservation system.

When the Saskatchewan organic farmers applied for certification as a class, they needed to demonstrate that there was a broad class of organic farmers who had been affected negatively by transgene flow from GE canola. One of the ways they did this was to submit scientific evidence that GE contamination was widespread on the prairies and in the seed supply. In an affidavit, Rene C. Van Acker, then an associate professor in the Department of Plant Science at the University of Manitoba, supplied evidence to support his opinion that

> because of canola seed's ability to remain dormant for four to five years, its ability to cross-pollinate, the common usage of it in crop rotations, the extensive use of GM canola, and GM contamination of seed lots, the likelihood of contamination of non-GM canola by GM canola in Western Canada is very high and perhaps absolute. (*Hoffman et al. v. Monsanto Canada Inc. et al.* 2002)[15]

In addition to Van Acker's expert opinion, the organic farmers believed they found a key piece of evidence when they succeeded in getting a study by the federal agriculture department, the AAFC, released, after much struggle, including a demand through the courts. The study, conducted by two scientists at the AAFC, one of whom was a vocal proponent of GE canola, found transgenes in samples of a wide variety of certified seeds—that is, seeds that are sold at a higher cost because they are supposed to meet certain standards of purity, which were not intended to have transgenic traits (Downey and Beckie n.d.).[16] In a press release, the organic farmers stated that the study would be crucial proof "because it provides scientific documentation of the widespread contamination that has all but wiped out the organic canola market" (Organic Agriculture Protection Fund 2002). These studies of the seed supply indicate the difficulty of obtaining non-GE seed, which is a requirement of organic production.

A ruling on the certification of the class action was issued in May 2005. Judge Gene Ann Smith determined that the Saskatchewan organic farmers did not have a cause of action regarding nuisance, trespass, or negligence. With respect to each of these, the judge did not think that there was a sufficiently direct connection between the biotechnology companies and the transgenes on the organic farmers' land to support the allegations. Each of these claims requires some kind of proximity or direct causation, and the judge chose not to interpret the law to fit these novel circumstances.[17] On the other hand, she affirmed that the plaintiffs' claims based on two provincial environmental laws were reasonable causes of action. She found that they did not apply to an identifiable class of people, though. In other words, individuals affected by transgenic contamination could bring lawsuits under these environmental laws, but a class action was not appropriate because the judge did not believe that Hoffman's and Beaudoin's experiences were widespread, or that GE contamination was necessarily the main reason why organic farmers were deciding not to grow canola.[18] In the proceedings, Monsanto Canada and Bayer CropScience had questioned the feasibility of growing organic canola along with the extent to which it was ever grown in Saskatchewan, and challenged the premise that GE canola created an obstacle to producing canola organically. An appeals court upheld the lower court decision in 2007, and the Supreme Court of Canada refused to hear a subsequent appeal.

Legal scholar Heather McLeod-Kilmurray (2007) offers a critical assessment of the judge's decision not to certify the farmers as a class. She indicates that the judge found no identifiable class "mainly because she focused a great deal of her analysis on the fact that although the farmers alleged environmental contamination, this was an individual experience for each farmer . . . [and] the individual issues would overpower the common ones" (ibid., 189). McLeod-Kilmurray notes that contrary to the judge's determination, "the risk of GMOs is not individual. They affect the ecosystem itself, the organic option itself" (ibid., 195). Furthermore, the class action procedure was created to resolve common issues efficiently and deter those who may cause widespread harm.

Why, then, were the organic farmers not allowed to proceed with the legal action? In part, McLeod-Kilmurray suggests, the decision may stem from the perception that the farmers are primarily seeking compensation for economic losses, which may appear to be individual rather than collective problems. "Bringing actions in tort law [such as negligence or nuisance] tends to obscure the nature of the problem involved in environmental cases and the full range of purposes the class seeks to achieve in bringing the litigation as a group" (ibid., 194). In this case, although "an essential part of what the farmers are seeking is greater oversight, participation, and accountability" in decisions about GE seeds, "once compensatory damages are sought, it seems that all the priorities, values, and goals of the private law come to the fore, as do the traditional roles of courts and citizens in private litigation" (ibid., 195). This is one of the pitfalls of framing the issue of GE contamination as primarily a "marketability" problem. It highlighted the individual economic losses more vividly than the collective grievances about GE crops, food production, and the power of the biotechnology industry.

Yet McLeod-Kilmurray concludes that the tendency to treat tort claims as individual matters is not the major source of the problem in the Hoffman case. Instead, she finds fault in the judge's interpretation of the law. "Part of the reason why the motions judge in *Hoffman* saw primarily individual issues of economic harm seems to be that she was particularly influenced by the idea that GM canola is 'safe.' . . . [T]he tone of the motions judge suggests a very clear attitude against a finding of liability against the manufacturers" (ibid., 196–197). Indeed, the ruling in this case protected the Canadian regulatory system from scrutiny.

Conflict surrounded the farmers' inclusion of an affidavit by Mae-Wan Ho, the director of the Institute of Science in Society, a UK-based organization that campaigns for social responsibility in science and offers critical perspectives on biotechnology. Ho presented evidence of the environmental and health risks of GE crops—which contradicted the assurances of safety offered by the biotech companies and the Canadian regulatory agencies. In response, Monsanto and Bayer urged the court to accept the Canadian regulatory approvals as proof of the safety of GE canola, and petitioned to have Ho's affidavit struck from the claim. The judge settled the dispute by saying that the question of whether GE crops are environmentally dangerous should be taken up at trial, not during the certification hearing. The judge's decision on this matter read, in part:

To allow this affidavit to stand would invite the defendants to file affidavits of expert opinion on the other side of this issue and risks seriously complicating and confusing the issues which are properly to be determined on the certification motion. In short, evidence as to whether or not GMOs are inherently environmentally dangerous, while relevant to the merits of the plaintiffs' claim, is not directly relevant to any of the issues to be determined on the certification application. (*Hoffman et al. v. Monsanto Canada Inc. et al.* 2006)

Thus, it would appear that the judge set aside matters of scientific controversy as irrelevant to her decision at this stage of the legal process, allowing them to be debated at trial.

Still, at other points in the ruling, the judge upheld the legitimacy of the government approval of GE canola, rather than leaving it to be debated during the trial. As the organic farmers pointed out in a subsequent appeal, there are signs throughout the ruling that Smith viewed the safety determinations of the Canadian regulatory agencies as uncontroversial and not open to dispute. The judge described Ho's scientific claims as "complicating and confusing the issues," while the assertions made in Canadian regulatory decisions ostensibly did not have that effect. As the plaintiffs mentioned in their appeal, the judge "devote[d] almost half her recitation of the facts in her certification decision to quoting (with evident approval) from the Canadian regulatory decision documents on the environmental safety of the Defendants' GMOs." The farmers went on to note, "Dr. Mae-Wan Ho's Affidavit specifically addressed the inadequacy of the Canadian regulatory regime to properly evaluate GMOs for safety" (*Hoffman et al. v. Monsanto Canada Inc. et al.* 2005a, 9). The appeal also remarked: "It is hardly judicial policy to at all times defer to government regulators. Everything from thalidomide to the exploding Pinto at one point received government regulatory approval" (ibid., schedule A, 3).

The court's ruling in this matter, however, in effect placed the science used in regulatory decisions beyond scrutiny.[19]

It was apparent in other ways that the judge took some of the safety assumptions for granted that the plaintiffs aimed to challenge through their legal action. For example, Smith used the terminology adventitious presence to describe the presence of transgenic material in unintended places. This is the preferred terminology of both the biotechnology industry and the Canadian regulatory agencies, and is strongly objected to by critics of GE crops because it is typically used to suggest that transgene flow is simply a benign commingling of materials. In this context, the judge's use of this terminology seemed to indicate an unexamined acceptance of the safety of approved GE crops.

Beyond the Courtroom

The courts turned out to be unwilling to question regulatory decisions about the safety of GE crops. Furthermore, and perhaps most frustrating for the farmers and other critics of the *Schmeiser* decision, the OAPF class action case failed to resolve the contradiction between the status of transgenes as intellectual property, on the one hand, and the absence of rules addressing the consequences of releasing those patented genes into the environment, on the other hand. When compared to the ruling in *Monsanto v. Schmeiser*, it is striking that the trial judge's findings indicated that the biotechnology companies could not be responsible for nuisance or other allegations because their actions were not the direct cause of the problems the farmers faced. In the case against Schmeiser, the distance between Monsanto and the Roundup Ready genes in the farmer's canola was vast—in terms of the steps that occurred between marketing the seed and its discovery on Schmeiser's farm—yet the company could still exert patent rights. The distance between the companies and the organic farms would seem to be no greater, but the law appears to accept this incongruity.

As in the previous chapter, then, it is worth asking whether legal mobilization is an effective strategy for the antibiotech movement. Legal mobilization can have different effects on the various types of participants in an activist network. It is evident that the experiences of organic farmers and other antibiotech activists surrounding the OAPF lawsuit were rather different. For antibiotech activists who were already committed to political mobilization, the class action may have been a drain on resources. Conversely, for organic farmers who were not actively involved in biotechnology politics before the lawsuit, the lawsuit was a gateway to engagement in further antibiotech struggles.

As legal scholars are well aware, "litigation is costly, slow, difficult to access, reactive, and can divert energies from concrete resolution of immediate environmental problems" (McLeod-Kilmurray 2007, 189). Given the obstacles that were encountered in the class action lawsuit, the antibiotech activists I spoke with expressed concern that the strategy of legal action was not helpful in building a movement, because it took time, leadership, and resources away from other types of activism. As one food and farming activist (Saskatoon, July 13, 2006) explained, when attention turned to the class action, other efforts suffered:

> Our case almost seemed to take the wind out of the sails of the other activists. . . . We had this biotech working group, and I was really involved, and we organized this conference, and we'd have meetings, and we'd do demonstrations and stuff like that. And then, eventually, when the lawsuit started . . . that's where I started putting my energy. . . . Somehow it seemed like the lawsuit was so big, and there's not that many people in Saskatchewan. It took almost everybody's attention that was working on GMOs. And some of the other stuff that might have happened if we hadn't done the lawsuit didn't happen. . . . Legal actions can suck up a lot of time and energy. . . . It's a different approach to try and solve a problem than building a political power and political base and mind-set and all that sort of thing.

Framing grievances in legal terms can cause social movements to lose their critical edge as well.

But legal struggles can also contribute to the mobilization of a social movement, particularly in articulating grievances and making rights-based claims. Indeed, for the organic farmers involved in the class action lawsuit, legal action led to a more explicitly political activism. The lawsuit generated connections to other struggles to defend non-GE agriculture, not only in Canada, but around the world too. Members of SOD and the OAPF have traveled to France, India, Denmark, England, Australia, New Zealand, Italy, and the United States to meet with antibiotech activists and speak about their experiences (Saskatchewan Organic Directorate 2006). Organic farmers also developed ties with the broader antibiotech movement in Canada. For example, SOD is now a member of CBAN, the major coordinator of antibiotech movement activity in Canada. Prominent antibiotech activists, such as David Suzuki, spoke out in defense of the Saskatchewan organic farmers and helped them raise funds to support their legal action.

For some, broader engagement has led to frustration with the state of the antibiotech movement in Canada. One organic farmer, for instance, said he had not previously thought of himself as an activist, but his involvement in the lawsuit had gradually turned him into one. He expressed frustration with the slow pace of the legal process and relative lack of public mobilization against GE crops in Canada. In part, his perception of the situation in Canada was based on his experience with the French antibiotech movement, which he encountered as a result of his involvement with the OAPF. He had been invited to give testimony in support of a group of people who were on trial for destroying a field of experimental transgenic corn in rural France. He said that the French defendants wanted

to use examples of what's happening in North America, where this has been going on for years, you know, where it's kind of open season, and the governments are favorable to it, and the civil society isn't mobilized enough, or able to, or too apathetic, or a combination of all the above, to really do something about it. But the climate in Europe is so, so different. I found that really exciting, to go there and participate in that.

He went on to describe a demonstration he joined in France, in which thousands of people marched to the town where the farmers accused of destroying the crops were on trial:

Walking into Grenoble at three thousand strong, with banners and singing, and the whole nine yards. And everybody was received really well. And I thought, wow, this is extraordinary! . . . So that sort of gave me, like, encouragement—you know, this is really exciting! And to come back here and it's like well, OK, you attend a couple of meetings and you do certain activities and that's great, but things just don't seem to be progressing. And that's partly to do with the way the justice system's set up, maybe. I don't know if there's more we can actually do. (organic farmer, Saskatchewan, July 15, 2006)

There have been few opportunities for farmers and antibiotech activists to challenge the way that biotechnology is governed in Canada. Regulatory agencies and the legislature have been unreceptive to grievances about genes out of place. Moreover, the OAPF found that the justice system was not "set up" to address the problems faced by organic farmers. So while participation in the class action might have led farmers to engage more deeply in political activism, it also revealed the shortage of political opportunities.

From Risk to Marketability

Since GE crops were first commercialized in the mid-1990s, government regulation in Canada has been narrowly confined to a small set of "safety" concerns. The government has actively promoted biotechnology, through patent law and the privatization of seed research, among other actions. Recent calls for government regulation based on the marketability of GE crops, though, suggest that despite the dominance of neoliberal ideas and so-called sound science approaches to regulation, the need for state intervention to protect agricultural producers is at least on the political agenda. By generating strong market demand for non-GE commodities, antibiotech activism and the organic movement have generated sizeable obstacles to the expansion of the agricultural biotechnology industry. In Canada, one of the major oppositional strategies has been to demonstrate that GE crops pose a threat to important export markets, and thus demand that seed producers withdraw GE seeds from the market. This strategy has succeeded when there is widespread agreement across dominant agricultural producers that key customers will not accept GE crops. The battle over Roundup Ready wheat, described at the beginning of this chapter, demonstrates that in a context where farmers are highly export dependent, opposition to GE foods in other countries can have a significant impact in producing countries.

Emphasizing consumer demands for non-GE crops sidelines the more fundamental critique of the current political economy of agriculture and the shortcomings of the regulatory system. Nevertheless, the vehemence of industry objections to proposals like Bill C-474, which would have expanded the criteria for deregulating GE seeds, indicates just how strongly the biotechnology industry wishes to maintain the current version of science-based regulation. Furthermore, as the OAPF lawsuit illustrated, making the case that GE crops pose a problem for marketing implicitly raises questions about their presumed safety and the trustworthiness of Canadian regulatory agencies. The Saskatchewan organic farmers' demands went well beyond protecting consumer choice. The farmers rejected the notion of coexistence between GE and non-GE crops—which would support a notion of consumer choice in the marketplace—in favor of a total ban on GE crops.

The struggle of organic canola farmers against the biotechnology industry, though, produced no tangible legal changes that would protect alternative agriculture in Canada. Indeed, over the years that the OAPF pursued the class action, the adoption of GE canola by Canadian farmers steadily increased—to over 90 percent adoption in 2009 (James 2009). The courts' rulings reflected the dominant perspective that genes out of place are merely a consumer preference issue, rather than an environmental problem, food safety concern, or any other matter that might require state intervention. The lawsuit also failed to establish any additional criteria for governing GE crops, such as marketability, economic impacts on organic farmers, or broader social concerns about the trajectory of technological change. Thus, instead of becoming objects of state governance, genes out of place remained unregulated in both the environment and canola seed supply.

7

Conclusion: Science and Struggles for Change

In a classic study of the social implications of technology, political philosopher Langdon Winner cautioned against the perils of contemporary discussions about risk. He observed that social critics and activists frequently turn to discourses about dangers to the body in order to stimulate popular protest. "It is clear," Winner (1986, 141) wrote, "that alarms about particular hazards will engage the public's imagination where more ambitious, general criticisms do not. Hence, the politics of hazards often becomes a strategic complement for or even an alternative to the politics of social justice." When the discourse of risk displaces demands for social justice, however, we uphold "the status quo of production and consumption in our industrial, market-oriented society." This is because "industrial practices acceptable in the past have become yardsticks for thinking about what will be acceptable now and in the future." Therefore, "the risk debate is one that certain kinds of social interests can expect to lose by the very act of entering" (ibid., 148–149). Indeed, defenders of new technological developments typically attempt to minimize public concern by making comparisons to other, accepted (but harmful) technologies.

Winner's analysis, though written before GE seeds were commercially available, is strikingly pertinent to the global conflict over GE crops. The antibiotech movement has gained broad popular support primarily by mobilizing fears about the risks of eating GE foods. Advocates of GE crops, in response, frequently indicate that critics are irrationally fearful of the risks of the technology, when compared to the known risks posed by other foods and agricultural practices. Often overlooked in this risk-centered debate about GE foods are the struggles of farmers, NGOs, and researchers committed to producing food in more sustainable and socially just ways. In this book, I have sought to foreground these commitments, focusing on the ways that antibiotech activism generates alternative ways to think about the implications of technology. Contrary to the prevalent idea that antibiotech activists are merely fearful of new technology, the case studies in this book revealed that objections to genes out of place stem from deep-rooted struggles for social change.

Farmers, agricultural researchers, and civil society organizations have worked for decades to create and promote alternatives to the dominant, industrial model of food production in many parts of the world. In Mexico, agronomists and agroecologists have collaborated closely with maize producers to improve native maize cultivation, conserve biological diversity, and build on traditional knowledge to create sustainable rural livelihoods. In Canada, farmers have defended their right to save seeds, and organic farming has grown rapidly as a more sustainable alternative to "chemical farming" and a way for smaller farms to remain economically viable. Farmers and activists in both countries questioned the notion that all technological change is progress, and that hazards can be anticipated and managed with existing regulatory tools. Their critiques did not focus solely on questions about risk but instead more fundamentally on questions of the following types: How should we distribute access to genetic resources? What technologies are useful for alternative food-provisioning systems and the rural poor? How much power should agribusiness companies have to shape the way we feed ourselves? Can we create systems of governance that promote sustainability as well as accommodate scientific uncertainty and ignorance?

In both countries, challengers had little influence over decisions made by regulatory agencies or the legislature regarding biotechnology. Activists thus sought change through other venues, demonstrating the diverse forms that challenges to dominant technologies can take. They externalized the conflict to international institutions, engaged in scientific research, sought change through the marketplace, and mobilized in the courts. In these concluding pages, I will discuss antibiotech activists' strategies of opposition along with what they indicate about the distribution of power and authority in an increasingly globalized economy and culture. In discussing these strategies, I will outline how conflicts over genes out of place in other parts of the world—and conflicts over technology more generally—are likely to resemble and differ from the cases analyzed here. Finally, I will consider what kinds of institutions might enable us to approach technological change differently.

Strategy, Science, and Power

By studying social movements, we learn about the complex ways that power is structured in a society. These cases show that even at a time of neoliberal restructuring of the relationship between states and markets, government regulation remains centrally important in defining the trajectory of agricultural biotechnology development and its alternatives. In this context, one way the biotechnology industry exerts power is by ensuring that the *criteria* that governments use to regulate GE crops preclude questions about their social desirability. A key obstacle in Mexico and Canada that challengers faced was the high expertise barriers to participation in national decisions about GE crops due to the official stance on science-based decision making. In some notable examples, farmers and antibiotech activists directly demanded that their governments take into consideration the cultural significance of maize and evaluate the marketability of GE crops. Networks of farmers and organizations in the antibiotech movement also sought change by targeting a variety of institutions and culture beyond the state. This is consistent with recent sociological research that draws attention to the overlapping, multilevel, and often contradictory institutions that make up a society as well as generate opportunities for social change. In struggles over the social consequences of genes out of place, challengers of GE crops highlighted institutional contradictions, such as tensions between domestic policies and international obligations. They used and combined institutional resources like laws and international treaties, scientific knowledge, organic standards, and elements of local culture in order to articulate new criteria for governing GE crops.

What lessons can be drawn from these examples? One conclusion is that cross-national differences in the status of GE crops do not stem only from features of distinctive national cultures, conflicts, and governance structures. The status of various GE crops in Mexico and Canada also reflects the position of each country with respect to international trade relationships and environmental treaties. The cultivation of GE maize is restricted in Mexico, for example, not only because of domestic opposition to it, but also because Mexican activists are able to appeal to the Cartagena Biosafety Protocol and other international bodies concerned with protecting biological diversity. Likewise, the widespread use of GE canola, but delay in commercializing GE wheat, reflects the differences in the global markets for those two crops.

These conflicts over GE crops reveal processes of movement mobilization that are broadly relevant, not only to contentious struggles over how to govern genes out of place, but also to other conflicts over technology and the environment. In these instances, farmers and activists used four main strategies: externalizing the struggle to international experts, carrying out civil society research, scrutinizing science in court, and using market-based tactics. Antibiotech activists are highly sensitive to ways that experts, key industries (such as food retailers), and environmental law exert power to shape both public policy and the behavior of the biotechnology industry. These strategies certainly do not exhaust the list of ways that social movements may challenge technology and question the political authority of science, but they do indicate the diversity of targets for movements that seek to create alternatives to the dominant technological trajectory.

Externalization to Experts

It is not uncommon for activist groups to call on counterexperts or use scientific studies as sources of leverage. Dissident scientists increasingly are allying themselves with social movements; they are also constructing alternative institutions such as "concerned scientist" NGOs. In the United States, organizations like Science for the People and the Union of Concerned Scientists formed in the 1960s and 1970s in order to provide scientists with ways to lend their expertise to political struggles (Moore 1996, 2008). Much more recently, in Mexico, a group of researchers created the UCCS, taking strong positions against GE maize and Mexican policies for biotechnology. The UCCS, much like the Union of Concerned Scientists in the United States, brings dissenting scientific perspectives into public view, undermining the government's claims that its policies are based on sound science. Experts were also a source of leverage in Canada. In the conflicts over GE canola, two Canadian researchers, Van Acker and Friesen, and one international concerned scientists' organization, the Institute of Science in Society, offered expert assessments that supported farmers in their conflicts with biotechnology companies.

I showed in chapter 3 that grassroots activists, transnational NGOs, international organizations, and scientific experts formed a new kind of relationship, which I termed an epistemic boomerang. An epistemic boomerang involves pressure from an organized group of scientific experts on a national state that is inattentive to local activists. Local antibiotech activists voiced their grievances to an international expert advisory group, which echoed those concerns back to the Mexican government in the form of a report and recommendations. This put pressure on the Mexican government to change its policies about GE maize, although it was significantly weakened by objections from the United States and Canada. Initially, activist appeals to the CEC framed genes out of place as a biosafety threat—a risk to the diversity of maize that could be assessed through scientific analysis. Antibiotech activists also unexpectedly opened up a space to articulate grievances about the ways that technological change, agricultural policies, and trade agreements threatened rural livelihoods and indigenous cultures. Externalization to the CEC thereby generated an opportunity to challenge the scientization of Mexican biotechnology politics by encouraging public participation and broadening the range of criteria for assessing the acceptability of GE crops. The CEC's organizational design made it particularly amenable to the epistemic boomerang process I described above. There are few, if any, other international organizations that are so explicitly dedicated to both providing expert recommendations and facilitating civil society input in regard to contentious matters. That said, the growing number of international institutions and transnational NGOs makes it increasingly likely that epistemic boomerangs will occur in other circumstances. NGOs are already frequent participants in many UN-organized technical meetings, including, for example, the Subsidiary Body for Scientific and Technological Advice of the UN Framework Convention on Climate Change (Miller 2001). The so-called participatory processes of other international organizations also hold the potential for mobilization that generates an epistemic boomerang. For instance, the World Bank holds public consultations on projects such as new hydroelectric dams. Although the participation that occurs in these hearings has been criticized (Goldman 2004; see also Cooke and Kothari 2001), encounters of this sort are likely settings for the kind of mobilization observed in the Mexican maize case.

Nevertheless, it is crucial to recognize that scientists often find their credibility is questioned when they take positions on politically contentious matters. This is likely in contexts where credibility is assumed to require distance from activist politics—which may vary across cultures and political regimes. Certainly, in the cases studied here, states and industries resisting the pressures of social movements benefited from perpetuating the notion that scientific advice must not be "tainted" by social concerns. Therefore, while activists may successfully gain the support of professional scientists, those allies face pressures to appear politically neutral.

Civil Society Research

Environmental monitoring, organized and carried out by activists and rural communities, was a second strategy of opposition to GE maize. An obvious target of these activities were Mexican government authorities that conceivably might have responded to reports of contamination with increased regulation of GE maize marketing and cultivation. There are examples in which civil society monitoring has had an effect on state policy and industry behavior. Greenpeace's discovery of unapproved

StarLink genes in food products led to a massive food recall (Taylor and Tick 2001), while in Costa Rica, government officials have voiced support for "community biovigilance" as a strategy to monitor transgenic fields (Pearson 2009, 731).

Beyond biotechnology, strategies of activist-led research and environmental monitoring are increasingly common. Several examples of activist-led environmental and health studies are well known, such as the health data collected by Lois Gibbs in the Love Canal crisis (Couch and Kroll-Smith 2000) and the medical expertise that AIDS treatment activists developed in the 1980s and 1990s. In the latter case, scientists recognized some AIDS treatment activists as relevant experts, accepting them into conferences and advisory panels, and taking their knowledge seriously—thus shaping AIDS research and treatment policies (Epstein 1996). Environmental justice struggles are other types of collective action in which citizens may gather environmental and health data in order to directly confront government agencies. These efforts are typically combined with collaborations with scientific allies who speak on behalf of affected citizens. Most recently, in the wake of the Fukushima nuclear disaster in Japan, ordinary citizens have organized themselves to monitor radiation levels in response to the failings of public officials to protect the public (Belson 2011).

The impacts of these types of efforts are varied. The findings of an activist-initiated study may help to prove that an environmental or public health problem exists when authorities have been reluctant to acknowledge it. Yet research produced by activists may not always be taken seriously in the institutional arenas where they are attempting to produce change. Furthermore, there may be resistance among participants in civil society research projects to conceptualize their work solely in terms that are legible to authorities, such as risk calculation. This was particularly evident in the Mexican maize case, when some activists began speculating that transgenes were causing maize deformities. State authorities and the scientific community were not always the targets for these claims. Distrust in the government, based on a legacy of disappointments and betrayals, led some maize activists in this diverse network to turn attention away from challenges to the state. Rural communities themselves became an increasingly important target of movement activities. Monitoring became a method of grassroots education, more than a means of gathering scientific evidence to support the policy positions of the antibiotech movement. This case indicates that while environmental monitoring may fail to provide political leverage, particularly when it leads community groups to make dubious claims, it might mobilize disadvantaged communities to defend their natural and cultural resources in other ways.

Scrutinizing Science in Court

Although the prevalence of litigation as part of a social movement strategy varies depending on the legal systems in place in particular countries, litigation may be used as a strategy to challenge a wide variety of technological developments, from defective breast implants to energy-producing dams. Legal mobilization is a prominent strategy of opposition to GE crops in Canada, the United States, and some other countries. As mentioned in an earlier chapter, farmers in the United States have demanded an injunction against Monsanto suing for patent infringement. In another recent case in Australia, an organic certification agency withdrew the certification of a farm after its fields were contaminated with transgenic canola seeds from a neighbor's field. The organic farmer sued the neighboring farmer in response (O'Brien 2011). In Canada, farmers and antibiotech activists confronted biotechnology companies in court, both as plaintiffs and defendants. In both of the cases involving GE canola, judicial rulings established the legal status of genes out of place, upholding the intellectual property rights of biotechnology companies and maintaining the status quo with respect to the regulation of environmental releases of GE crops. Moreover, in both cases, the proceedings became highly scientized—that is, scientific evidence and expert witnesses were central to the courts' rulings.

Close scrutiny of the legal proceedings suggests two conclusions about legal mobilization in defense of non-GE agriculture. First, unsurprisingly, to the extent that biotechnology companies have more access to high-status scientific experts, they have an advantage in a courtroom that narrowly interprets questions about biotechnology as technical matters. Second, however, judges have great latitude in deciding how to construct analogies between genes out of place and existing law. Although this clearly did not happen in these cases, it is not difficult to imagine a situation (akin to the Harvard mouse case) in which broad public concern about GE crops might shift the ways that judges interpret the law. In litigation involving GE alfalfa, for example, the courts ruled that existing environmental law had to be applied to the controversial new crop, requiring a much more detailed environmental impact statement than is typically carried out for GE crops in the United States. It made use of a frequently overlooked requirement of the National Environmental Policy Act, an assessment of the potential socioeconomic impacts. Another lesson to be drawn from these cases is that patent litigation is one surprising way that social movements may mobilize for change. Much as criminal defense trials are often part of social movements involving civil disobedience (Barkan 2006), patent infringement cases (involving genes, software, or other controversial intellectual property) potentially generate a new venue for protest, although they also may diminish the momentum of a social movement by draining resources and harassing farmers.

Science and technology can be opened up to public scrutiny in courtrooms, but science can be an unreliable ally to those who seek to challenge the status quo. It is a friend to the antibiotech movement to the extent that people need knowledge about the changes happening in the natural world. In the courtroom, counterexperts can challenge claims about the safety, precision, and controllability of GE. More generally, competing expert witnesses expose the degree of uncertainty and range of interpretation of scientific evidence, and lawsuits may reveal broader debates about the social consequences of technological developments. On the other hand, legal rulings may generate scientific closure prematurely, and it is difficult to predict how a judge or jury will interpret conflicting knowledge claims. As in the organic farmers' class action, for instance, a judge may choose to defer to existing regulatory science rather than consider contradictory knowledge. Furthermore, legal action is costly and can divert attention from other actions.

Consumer-Oriented Strategies

A fourth strategy of opposition to GE crops is to develop product labels and alternative markets, mobilizing consumers, food processors, and retailers to reject GE foods. Worldwide, consumer activism has made a significant impact on the biotechnology industry, by both influencing government regulation and driving food processors and retailers to reject GE foods. Consumer opposition, in other words, is not limited to practicing so-called choice in the marketplace. Just as consumers can act collectively to demand food safety regulations, farmers can act collectively to make demands on the seed industry, responding to the pressures they face from consumers (through food processors and retailers) in export markets.

Market-based political activism is not new. Social movements have historically used a variety of market-based protest strategies to produce social change, including boycotts, the formation of consumer cooperatives, and the creation of alternative markets and labels (such as fair-trade or sweatshop-free goods). From abolitionist refusal to purchase slave-produced goods to present-day boycotts of rain forest timber, social movements have used consumer action to protest and disrupt the practices of corporations that are perceived to be contributing to a social problem. Successful boycott campaigns of the past were usually part of a larger set of tactics to achieve political change, especially protections for workers.

Today, though, many transnational boycott campaigns attempt to bypass the national state, opting for new institutions representing civil society and private monitoring programs (Seidman 2007). Organic food standards and labels generally align with this trend, calling on consumers to "vote with their dollars" rather than press for stronger regulation on agrichemicals and environmentally destructive farming practices. Still, some research indicates that consumer opposition to GE food can lead to changes in the way the state governs agricultural biotechnology. European consumer rejection of GE foods, for instance, not only affected the practices of food companies; it was also a crucial factor in persuading states to create tougher regulations of GE crops (Schurman 2004).

In the cases of organic canola and wheat in Canada, farmers not only participated in alternative markets and pressured seed companies to change their practices; they called on the state to take action as well. In moving from the marketplace to the state, however, organic farmers encountered new expertise barriers and formal criteria for science-based decisions. They therefore also found themselves calling on the state to expand its criteria for regulating new technology. Still, the market orientation of the organic and antibiotech movements remained in the forefront, as opponents of GE crops primarily highlighted the problem of marketability. Activists in Canada have continued to fight for government regulation based on the marketability of GE crops, and concerns about marketability are driving government restrictions on GE crops in other parts of the world, too (Paarlberg 2002).

However, when antibiotech activists, who typically have a wide range of social and ecological concerns about GE crops, frame the issue in terms of marketability, there are several potential pitfalls. The emphasis on serving customers diminishes the transformative potential of alternative agriculture movements, turning attention from questions about "how we, as a society, want to live" toward "what we, as consumers, want to eat." Furthermore, assertions about marketability depend on either the demands of distant consumers or the regulatory status of GE crops in foreign countries, and opponents of GE crops cannot count on restrictions in export markets to drive changes in the governance of biotechnology. Willing or unknowing consumers may always be found in a different national market, and organic farmers have had less success when their opposition to GE crops goes against the dominant agricultural interests. Specifically, in the case of canola, conventional farmers have eagerly adopted herbicide-tolerant GE crops. The seed industry made sure that importing countries, particularly Japan, would accept transgenic canola before the identity-preservation system was dismantled and GE canola was widely introduced into Canadian agriculture. It appears that Monsanto is attempting to do the same with Roundup Ready wheat: securing agreements with importing countries in order to show that there is sufficient demand for GE wheat produced in Canada.

Notwithstanding these pitfalls, mobilization that uses market institutions may still be a viable strategy for challenging the trajectory of agricultural biotechnology. In comparison to more sweeping critiques of the biotechnology industry, the regulatory system, and the imbalance of power in the food system, demanding protection of markets for non-GE crops is a fairly conservative call for reform. In Canada, however, it offered a way to challenge the criteria used in governing GE crops, moving beyond the narrow focus on scientific risk assessment.

New Criteria, New Experts

Taken together, challenges to GE crops indicate that regulatory conventions that privilege science-based claims are highly resistant to change. But it is not unheard of for governments to recognize a broader range of criteria for permitting the commercialization GE crops or evaluating the impacts of other industrial developments. Indeed, there are numerous examples in which social and economic impact assessments have been built into regulatory frameworks. In the United States, socioeconomic analysis is required as part of the environmental impact assessment process under the National Environmental Policy Act of 1969. Social scientists developed a large body of literature about the methods and procedures for carrying out social impact assessments under this law (Freudenburg 1986; Burdge 2003; Lockie 2001; Vanclay 2006). In the 1990s, social impact assessments "fell into disuse" in the United States (Turnley 2002, 2), but the idea has been adopted in other parts of the world. The Cartagena Protocol for Biosafety indicates that socioeconomic considerations—particularly the impacts of biodiversity loss on indigenous cultures—may be relevant to decisions about trade in biotechnology (Stabinsky 2000; Kleinman and Kinchy 2007). European regulators have also considered a variety of strategies related to the social acceptability of technology. In the European Union, formal proposals were made in the 1980s and 1990s to carry out assessments of the socioeconomic impacts of agricultural biotechnology, such as compatibility with existing market regulations.

These were never approved (Kleinman and Kinchy 2003), but now the European Union is moving toward a new approach to the regulation of GE crops that would allow member states to ban the technology if it is deemed socially undesirable, even if granted safety approvals at the EU level (Alapekkala 2011).

Efforts to include socioeconomic considerations in environmental and biotechnology governance are not necessarily inclusive of public concerns about new technologies. In fact, socioeconomic assessment can also be scientized, quantified, and left to experts to assess. Yet beyond expanding the types of issues considered in regulating industrial developments, there is growing recognition of the value of public participation and deliberation about technological change. A variety of scholars have enthusiastically described examples of participatory processes as models for reinvigorating democratic values in the face of scientized politics (see, for example, Fischer 2000; Maasen and Weingart 2005). There are signs that science and technology are being opened up to public scrutiny in new ways. Since the 1970s, governments and international organizations have experimented with a variety of mechanisms, such as participatory technology assessment procedures, consensus conferences, and "citizen science" projects, to generate a role for ordinary citizens in the evaluation of science and technology. Such formal participatory processes are becoming more prevalent worldwide, leading some scholars to describe a "participatory turn" in science and technology policymaking (Jasanoff 2003, 235).

It is clear, however, that the extent and nature of public participation varies widely across countries, and that national or local governments may be reluctant to increase public participation in decisions that matter. In response to the GE maize scandal, it was not the Mexican government or Mexican research institutions that engaged the public but rather a trinational advisory body, the CEC, which ultimately had little influence. Furthermore, as critical observers have pointed out, in many cases, participatory processes are too narrowly constrained. In Canada, for instance, efforts to gather public input on biotechnology policy were boycotted by antibiotech activists, who viewed the process as merely a way to legitimize predetermined outcomes (Prudham and Morris 2006). The activists' perspective was likely well founded, given a number of studies of participatory processes that offered only a limited terrain for debate about the implications of technology, focusing primarily on the identification and management of risks while avoiding more fundamental questions about the kind of society we want to live in (Levidow 1998; Bereano 1997).

Some observers caution that participation is often deployed as strategy of political elites, co-opting social movements and lay knowledge as well as guiding people to conduct themselves in ways that do not challenge dominant relations of power (Hess 2007; Agrawal 2005).

Despite their flaws, existing policy ideas of social impact assessment and participatory forms of governance are important steps toward restructuring the governance of technology in ways that are explicitly responsive to social priorities. The struggles over GE maize and canola show that a model of scientific risk assessment that excludes questions about the desirability of technological change is a major obstacle to the survival of alternative forms of agriculture. The persistence of social conflict over GE crops strongly suggests that it is time for new "rules of the game" for the governance of technology, to reassess what is accomplished when we rule out considerations that are not easily categorized as scientific. Decisions about technology are, after all, decisions about how we want to live. It will not be easy to change deeply entrenched ideas and established rules about the appropriate ways to govern technology, and it is unlikely without a sustained, organized challenge. But changing the terms of debate is essential to the pursuit of more sustainable, socially just forms of agriculture. Anything but the status quo will be impossible so long as we collectively fail to regard the politics of technology as a matter of public concern.

Notes

Chapter 1

1. In this book, I use the terms genetically engineered and transgenic rather than genetically modified in order to make a distinction between the new techniques of biotechnology and earlier methods of modifying the genetic makeup of a plant population through other breeding methods.

2. I discuss the scientization of biotechnology politics and theoretical underpinnings of this idea more fully in chapter 2. A variety of scholars have used this concept in relation to biotechnology politics, most extensively Les Levidow, who, with a variety of colleagues, has thoughtfully examined the evolution of regulatory policy in Europe since the 1990s (Levidow and Carr 1997; Levidow 1998, 1999, 2001, 2002, 2003; Levidow, Murphy, and Carr 2007; Murphy and Levidow 2006). Other scholars, observing the way that GE crops are debated publicly, have used variations on the scientization notion. Anthropologist Chaia Heller (2001, 2004), for example, notes the "riskification" of biotech politics in France, while sociologist Fred Buttel (2005) refers to the "environmentalization" of biotechnology politics in the United States.

3. I borrow this phrase from Javier Lezaun (2004), who, like me, takes inspiration for the concept from Mary Douglas ([1966] 1978).

4. In this book, I refer to participants in a social movement as "activists" or "challengers." When possible, I try to be as specific as possible about the kinds of solidarities that particular activists express. For example, I may refer to someone as an "antibiotech activist," "indigenous rights activist," or "environmental activist" because of the kinds of organizations they belong to as well as their political activities prior to the anti-biotech movement.

5. For a detailed history of the regulatory decisions and legal challenges surrounding Roundup Ready alfalfa, see Straka 2010.

6. The role of the WTO in institutionalizing scientific risk assessment as the sole basis for regulating GE crops is discussed in chapter 2.

7. For a detailed look at the assertions made about GE crops and world hunger, see the published debate between Miguel Altieri and Peter Rosset (1999a, 1999b) and Martina McLoughlin (1999).

8. Although there is not much variety in the GE traits that are currently in commercial cultivation, investments are being made in research that seeks to develop drought tolerance in major crops, increase crop productivity, improve nutritional content, and solve other problems that farmers face worldwide (Castle, Wu, and McElroy 2006). At the same time, GE plants have also been put to industrial purposes not meant for human consumption. Some companies, for example, have developed plants that produce pharmaceutical chemicals (Kaiser 2008).

9. The US Department of Justice announced in 2009 that it was investigating anticompetitive conduct in the seed industry.

10. In the 1980s and early 1990s, several rural sociologists, drawing on Marxist theoretical traditions in the study of agriculture, anticipated that the new biotechnologies would increase corporate control over food production and disempower farmers (Kloppenburg [1988] 2005; Busch 1991; Goodman, Sorj, and Wilkinson 1987; Buttel 1995; Buttel et al. 1984; Kenney and Buttel 1985; Buttel and Belsky 1987). For a review of these studies and more recent social science perspectives on biotechnology, see Goodman 2003.

11. In the FDA's case, regulators approach the question of food safety by reviewing evidence that developers of GE foods gather in order to show that the new product is "substantially equivalent" to conventional foods. If substantial equivalence is demonstrated, the FDA regards the GE food as safe and exempt from further regulatory scrutiny. For an analysis of the failings of this system, see Pelletier 2005, 2006.

12. It is also worth pointing out that some environmental impacts are experienced directly by farmers. Key problems affecting farmers are the development of weeds that are resistant to the heavily used Roundup herbicide and insect pests that can withstand pesticide-producing plants. Just as in the use of traditional chemical herbicides and pesticides, eventually a pest population evolves resistance, forcing farmers to find a new way of controlling them. This is particularly problematic when "gene stacking" occurs—that is, when multiple transgenic traits appear in one plant, due to the cross-pollination of different types of transgenic plants. For example, canola can become a troublesome weed when it cannot be killed by multiple herbicides (Hall 2000).

13. In the 1980s, a small number of plant scientists like Norman Ellstrand (2003), an expert on plant gene flow, began to consider whether gene flow from transgenic crops could pose new environmental risks. Nevertheless, in the late 1980s and early 1990s, transgene flow was not expected to be significant, and scientists and regulators tended to underestimate the frequency of gene flow from transgenic crops (Andow and Zwahlen 2006). Since then, scientific attention to transgene flow has intensified, partly as a consequence of a number of high-profile cases in which transgenic material was discovered where it was not intended.

14. Some more sophisticated advocates of GE crops recognize that ignorance is not fully to blame for the opposition. For example, Robert Paarlberg's (2008) recent book, which argues that opposition to GE crops in wealthy countries has stalled the development of GE crops for agriculture, recognizes that opposition cannot be blamed "on public fears or misunderstanding of science" (p. 36). This

point was overlooked in the foreword by Borlaug and Carter, quoted above. Paarlberg argues that "citizens in rich countries dislike GM [genetically modified] foods and crops not because those products carry new risks but instead because they so far have provided consumers with no new benefit (p. 32). However, Paarlberg dismisses some other key explanations for public opposition to GE crops, such as concerns about the power of multinational corporations and the expansion of intellectual property rights to products of biotechnology. As evidence, he points to the general acceptance of biotechnology in the development of medical drugs, despite the involvement of multinationals and patents in that area of research. The relevant difference between biomedical biotechnology and GE crops, which Paarlberg overlooks, is that the latter has been adopted as a key issue for a large and long-standing network of advocacy organizations addressing issues of agriculture, hunger, and plant genetic resources. There is no comparable advocacy network dealing with issues of medical research that has taken a strong position on biotechnology in drug research and put the issue on agenda for public debate. In other works, Paarlberg ignores the role of social movements in shaping public opinion about technologies and their applications in particular fields.

15. Martha Herbert (2005, 59–60), a medical researcher and critic of agricultural biotechnology, indicates that proponents of GE crops may themselves lack a full understanding of the issues at stake:

Proponents of GE foods do not appear willing to engage in open and transparent debate. We may attribute some of this to vested interests, but that does not fully explain the problem. Many GE food proponents not only fail to address the concerns of GE food critics but appear unable even to comprehend the criticisms. They frequently claim that they themselves (the proponents) are uniquely "scientific" and their critics merely "emotional." Sometimes this rhetorical strategy is a disingenuous public relations maneuver. But it also reflects genuine naivete. Arguments about GE food's threats to organismic, ecological, and cultural complexity and diversity may simply be incomprehensible to many GE enthusiasts.

16. In advancing this approach to the study of social movements, Armstrong and Bernstein draw on and synthesize a variety of critiques of the "political process model," a dominant theory among US scholars of social movements since the 1980s (McAdam 1982; Tilly 1978; McAdam, Tarrow, and Tilly 2001). The political process model holds that social movements are rational attempts by groups excluded from political power to advance collective interests through noninstitutionalized means. Research in this tradition has been enormously insightful; however, as Armstrong and Bernstein summarize, over the years a number of substantial critiques have been amassed. Sociologists have called into question the state centeredness of the theory, which seems to ignore some social movements, and incorrectly assume that the state is the sole or primary source of power. Furthermore, political process theory has struggled to address questions of culture and identity in movements, and does not adequately consider the construction of grievances and motivations to act. Armstrong and Bernstein also set their approach apart from European "new social movements" (NSM) theory, which emphasizes questions of identity, meaning construction, and cultural transformation (Melucci 1996; Offe 1985; Touraine 1981). NSM theory offers an appealing

framework for understanding activism that targets institutions besides the state (such as the nuclear power industry) and the emergence of "expert" identities among laypeople. The shortcoming of NSM theory, however, is that it posits that the "new" movements such as environmentalism and health movements stem from macrostructural shifts toward a postindustrial society. Such bold claims about the emergence of a new social order do not hold up to empirical scrutiny, particularly with respect to movements of the Global South.

17. Commenting more generally on the impacts of planting GE crops, Greenpeace International (2007) indicates that GE organisms "can spread through nature and interbreed with natural organisms, thereby contaminating non 'GE' environments and future generations in an unforeseeable and uncontrollable way. Their release is 'genetic pollution' and is a major threat because GMOs cannot be recalled once released into the environment."

18. Kathleen McAfee (2003, 2008) has published a number of important articles addressing the Mexican maize controversy and other GE crop issues. Her observations that anti-biotech protests in Mexico reveal an alternative vision of agriculture and rural life, and that technoscientific debates about contamination lose sight of the broader consequences, influenced and resonated with my own interpretation of this case. Other studies of the Mexican maize controversy also contributed to my understanding of this case, including work by Jason Delborne (2005, 2008), Elizabeth Fitting (2006a, 2006b), Joel Wainwright and Kristin Mercer (2011), Gabriela Pechlaner and Gerardo Otero (2008), and Manuel Poitras (2008a, 2008b).

19. I am not the first to apply Douglas's insights about pollution and purity to the topic of GE crops. Lezaun (2004) and Michael Carolan (2008) have both cited Douglas to make related points. In addition, building on Douglas's ideas, Martijntje Smits (2006) has developed a "monster theory" to address modern perceptions of technological risk. She argues that many new technologies violate accepted categories. For example, she interprets plastics as an "ambiguous substance that did not fit into the cultural scheme," thus leading to strategies to "cope with plastic waste monstrosity, such as material recycling, converting waste into new primary products," and the like (ibid., 498). In a typology of approaches to dealing with new technologies, Smits uses action groups that destroy fields of GE crops as an example of "monster exorcists," those who want to expel the monster because it "simply does not fit within the symbolic order as it is meant to be" (ibid., 500). While there is much value in Smits's insights, I am not using Douglas's ideas in quite the same way here. Rather than emphasizing the ways that GE crops challenge already-fixed cultural categories, I am more interested in how activists and their opponents actively struggle over the social order in which the new technology fits (or is "out of place").

20. It has been argued that governments around the world have aligned their biotechnology policies either with the United States or European Union, depending on their trading ties (Bernauer 2003). The United States certainly has had enormous influence on public policy for GE crops in Canada and Mexico, not least through the North American Free Trade Agreement, but the United States is not the only influence. In Canada, for instance, GE contamination of organic canola

destroys the marketability of that crop in the European Union, where consumers and regulators have demanded non-GE foods. Concerns about marketability in international markets have prompted Canadian farmer groups to question the desirability of other transgenic crops too, such as Roundup Ready wheat (Eaton 2009). In Mexico, debates about governing GE crops in that country are strongly influenced by international environmental agreements that place special emphasis on protecting areas of high biodiversity (Gupta and Falkner 2006).

21. I interviewed fifty-seven informants related to the GE maize controversy, including twelve scientists, nine government officials, and thirty-six maize activists (ten of whom were maize producers or lived in rural communities). I interviewed twenty-five informants related to the GE canola controversies, including thirteen opponents of GE canola (eleven of whom were farmers), four scientists, and an assortment of representatives of government agencies, industry groups, and the Canadian Wheat Board.

Chapter 2

1. Rural sociologists have extensively analyzed these changes, generating a variety of theoretical perspectives on the globalization of agrifood systems through the growth of transnational agribusiness and global commodity chains (Goodman and Watts 1997; McMichael 1995; Bonanno et al. 1994; Higgins and Lawrence 2005).

2. These international bodies include Codex Alimentarius for food safety standards, the International Plant Protection Convention for plant health standards, and the Office of International Epizooties for animal health standards.

3. The text of the Cartagena Protocol on Biosafety is available at http://bch.cbd. int/protocol/text. The concept of biosafety took shape in international negotiations over the UN Convention on Biological Diversity in the early 1990s (Escobar 1998). The idea of biodiversity reflects efforts by conservation biologists and international environmental organizations to promote wildlife conservation and the protection of endangered species while "maintain[ing] an aura of scientific respectability" (Takacs 1996, 76). It offers the promise of calculability, encouraging the quantification of genes, species, and ecosystems, which are frequently assigned a market value as "biological resources." In this context, biosafety refers to the "safe handling" of GE organisms so as to avoid negative impacts on biodiversity.

4. Anthropologist Arturo Escobar (1998) points out that although there is much to criticize about the market logic and processes of rationalization underlying the idea of biodiversity, social movements that are confronting environmental destruction have found it productive to appeal to biodiversity discourses and institutions. These efforts create opportunities to articulate alternative views of biodiversity conservation from the perspectives of oppressed groups.

5. Thomas Bernauer (2003) uses the phrase "regulatory polarization" to describe the differences between US and EU approaches to agricultural biotechnology, indicating that the two governments are at extreme odds when it comes to regulat-

ing GE crops. While the European Union has restricted the use of GE crops, the United States has taken a much more permissive stance.

6. Here, Wynne echoes Ulrich Beck (1992, 58), who argues that risk statements always, unavoidably, "contain statements of the type *that is how we want to live*." To take the present case, statements that there is a low risk of harm from GE crops implicitly say, "We want or accept a system of food production in which multinational corporations sell patented, genetically homogeneous seeds to farmers around the globe." Conversely, statements that GE crops are highly risky often implicitly mean, "We do not want that kind of agricultural system, so any amount of risk is unacceptable." Frequently, statements of preference about the social order remain unstated, expressed instead in the scientized language of risk assessment and management.

7. The moratorium was then lifted in June 2003, in order to permit experiments dealing with transgene flow (Secretariat of the Commission for Environmental Cooperation 2004). Applications for GE maize commercialization were still not accepted at that time.

8. The English translation of the manifesto, "Defender nuestro maíz, cuidar la vida," is available at http://weblog.greenpeace.org/ge/archives/Oaxaca%20MAN IFIESTO.pdf.

9. For more about the history of canola, see Kneen 1992; Tanaka, Juska, and Busch 1999; Busch et al. 1994; Busch and Juska 1997; Juska and Busch 1994.

10. For more detail about the Canadian regulatory system for GE crops, see Abergel and Barrett 2002; Barrett and Abergel 2000, 2002.

Chapter 3

1. The epistemic boomerang is also different from the model of transnational advocacy in that it is not necessarily transnational. Still, because of the transnational nature of many scientific communities and supporting role of international organizations, it is likely that an epistemic boomerang will cross national borders.

2. Some examples illustrate this point. Sylvia Noble Tesh (2000) describes a mid-1970s' case in which residents of Alsea, Oregon, organized against the US Forest Service's spraying of the herbicide 2,4,5-T, believing it to cause health problems. One resident, a schoolteacher, collected community-level data that seemed to indicate a correlation between the herbicide spraying and the incidence of miscarriage. Friends of the Earth, an environmental advocacy NGO, publicized the teacher's study and pressured the US Environmental Protection Agency (EPA) to conduct a full scientific inquiry. Because the EPA was already attuned to the issue and had recently asked manufacturers to produce research showing that the herbicide was safe, the schoolteacher's amateur epidemiological study "landed in a fertile field" (Tesh 2000, 15). The EPA created a study team to more fully examine the link with miscarriage. Ultimately, citing the research carried out by its study team, the EPA suspended the herbicide's use. In this case, the presentation of the community's concerns in the form of an epidemiological study (as opposed to, say, a critique of the chemical industry) successfully bridged the local group's concerns

and the perspectives of EPA's toxicologists. This case also indicates the important role of NGOs in connecting local groups with experts.

3. Chapela had earlier aided the organization and the communities it worked with in negotiating a bioprospecting agreement with Sandoz (later Novartis), since he had once worked for that company.

4. Other organizations joined RAFI in the debate about "the seeds issue" in the 1980s, including Friends of the Earth, International Organization of Consumers Unions, and the Pesticide Action Network (Kloppenburg [1988] 2005). All of these NGOs are significant players in international biotech politics today. Building on the issue of genetic erosion, many of these activists are strongly opposed to the introduction of GE crops into places considered to be centers of origin and genetic diversity.

5. Since 2001, the Via Campesina network has focused on GE crops and biodiversity as one of its primary issues of concern (Desmarais 2002). A 2002 statement called for "no patents on life and a moratorium on the genetically modified crops which lead to the genetic pollution of essential genetic diversity of plants and animals" (NSO/CSO Forum for Food Sovereignty 2002). Many Via Campesina organizations, such as in Mexico and the NFU in Canada, are prominent critics of genetic engineering in their own countries and also join protests at the international level, such as demonstrations during the WTO meetings that took place in Seattle and Cancún.

6. Agroecology is defined in many different ways. It is a science that focuses on the interactions between plants, animals, humans, and the environment in agricultural systems. It is also a movement to promote agroecological practices. See, for example, Cohn et al. 2006; Altieri 2004.

7. Cited from an open letter to US president Barack Obama on the occasion of his visit to Mexico.

8. In Latin America, organized indigenous rights movements have emerged in the past two decades, largely in response to political-economic transformations. Deborah Yashar (1998) explains that with neoliberal economic reforms, indigenous communities lost the community autonomy and corporatist claims on the state that they previously enjoyed, albeit under a less democratic state. Building on social networks through churches and peasant unions, indigenous groups mobilized around the issue of land rights "as a means to achieve material survival with local political autonomy" (ibid., 38). Indigenous activists in Latin American make claims on the state on the basis of their indigenous identity when state reforms have threatened the community spaces in which indigenous practices and authority had been institutionalized. In the Mexican case, the Salinas administration's decision to dismantle constitutional protection for communally held ejido land—where communities previously had a certain degree of autonomy and security—was a key factor motivating indigenous communities to protest (Harvey 1998). Also contributing to indigenous activism in the early 1990s was symbolism of the anniversary of Columbus's "discovery" of the Americas and indigenous peoples' five hundred years of resistance. The anniversary "gave an unprecedented boost to indigenous social mobilization, spawning new grassroots organizations, rein-

vigorating existing organizations and pan-Indian ties, and heightening international awareness of the indigenous perspective on that history" (Carruthers 1996, 1015).

9. Ironically, the discovery was made in the community laboratory created as part of the agreement with Sandoz, under the guidance of Chapela, who helped to negotiate the Sandoz agreement. See note 3.

10. UNESCO (http://www.unesco.org/culture/ich/en/RL/00400) describes traditional Mexican food as "a comprehensive cultural model comprising farming, ritual practices, age-old skills, culinary techniques and ancestral community customs and manners. It is made possible by collective participation in the entire traditional food chain: from planting and harvesting to cooking and eating. The basis of the system is founded on corn, beans and chili; unique farming methods such as milpas (rotating swidden fields of corn and other crops) and chinampas (man-made farming islets in lake areas); cooking processes such as nixtamalization (lime-hulling maize, which increases its nutritional value); and singular utensils including grinding stones and stone mortars. Native ingredients such as varieties of tomatoes, squashes, avocados, cocoa and vanilla augment the basic staples."

11. Details about this workshop and manifesto are available at the GEA Web site: http://www.geaac.org.

Chapter 4

1. For more on the history of the idea of biodiversity, see Takacs 1996; Farnham 2007. For discussions of biodiversity governance in Mexico, see Brand and Gorg 2003; Hayden 2003.

2. For an excellent summary of the research on transgenes in Mexican maize, see Wainwright and Mercer 2011.

3. NGOs, advocacy groups, groups of volunteers, or other civil society organizations sometimes carry out scientific research themselves when they are faced with an unmet need for knowledge, or "undone science" (Woodhouse et al. 2002). Undone science refers to the "areas of research that are left unfunded, incomplete, or generally ignored but that social movements or civil society organizations often identify as worthy of more research" (Frickel et al. 2010). Activists may see particular questions as worthy of more research for a combination of reasons. They may have questions about the causes of their own health problems or impacts of industrial developments on their local environments, but find that scientific research on the invisible hazards that concern them has never been done. They may also find that without scientific data, they have little political influence. When regulatory decision making is scientized, those who fail to frame their positions in an appropriately scientific way run the risk of being excluded from the decision-making process. In this context, social movements find they need to gather scientific evidence to support their opposition to particular industries or technological developments. Civil society research can be an effective response to the problem of "undone" science because activists and ordinary citizens may produce knowl-

edge about environmental and health hazards that certified experts are unable or unwilling to recognize.

4. In his original formulation, for Beck (1992) the risk society is a period of modernization that emerges after a society is wealthy and technologically advanced enough that it is no longer focused on questions about the distribution of scarce goods. "Wealth-distributing" societies (modern welfare states) are transforming into "risk-distributing" societies. Beck refers to this transition as "reflexive modernization," in which science and technology are applied to critiquing as well as managing science and technology, in a continuously reflexive process. Yet it is now evident that the reflexivity that Beck describes, which opens science and technology up to scrutiny, is occurring in countries around the world, regardless of the state of political-economic development. In his later work, therefore, Beck (1999, 2) refers to a world risk society, indicating that "non-Western societies share with the West not only the same space and time but also—more importantly—the same basic challenges of the second modernity [i.e., reflexive modernity] (in different places and with different cultural perceptions)."

5. The results of the INE study were never published in a peer-reviewed journal, but were discussed in the press and appeared in conference proceedings (Ezcurra and Soberón Mainero 2002).

6. StarLink corn, sold by the biotechnology company Aventis, was approved only for feeding livestock in the United States because of indications that it could cause allergic reactions. Nevertheless, it was discovered in food products in 1999, leading Aventis to recall the StarLink corn varieties (Taylor and Tick 2001).

7. The research methods of Ortiz-García and her colleagues have been closely scrutinized. A reanalysis of the data found that the major conclusions were unjustified (Cleveland et al. 2005). One critique characterized the samples analyzed in the study as "unrepresentative" and the statistical analysis as "inconclusive" because of the sampling methods (Soleri, Cleveland, and Cuevas 2006). Subsequent studies, using different methods of sampling and molecular analysis, have found transgenes in maize landraces (Dyer et al. 2009; Piñeyro-Nelson et al. 2009a, 2009b).

8. False negatives are indeed an important concern among scientists studying transgene flow, but they are generally regarded as rare occurrences (Piñeyro-Nelson et al. 2009a, 2009b).

9. The literature on citizen science indicates that having scientists take an interest in building on activist-initiated research is essential to transforming activist-produced data into the desired political outcomes. For instance, in one well-known case, the residents of Woburn, Massachusetts, collected research data establishing a relationship between toxic waste in the water supply and the occurrence of childhood leukemia. This study enabled the affected families to pursue a lawsuit against the companies responsible for dumping the toxic waste and obtain a sizable out-of-court settlement. This example of "popular epidemiology" is widely cited as a prototype for successful citizen-initiated participatory science (Brown and Mikkelsen 1990). Key to the citizens' success was their collaboration with Harvard School of Public Health scientists, who took an interest in the data that

community activists were collecting, and helped them to design and carry out a more detailed public health study. Similarly, an activist-initiated asthma study in West Harlem eventually led to a scientific collaboration with researchers from Columbia University. The findings of the collaborative research project were the impetus for important environmental policy initiatives (Corburn 2005; Brown et al. 2003). In other examples, scientists and activists do not collaborate, but regulatory agencies develop studies inspired by research done by ordinary citizens (Tesh 2000, 15–16).

Chapter 5

1. In addition to innumerable references to this case in the popular media and debates about GE crops around the world, there are a growing number of scholarly analyses of this case and its implications (Bereano and Phillipson 2004; Busch 2002; Clark 2004; Garforth 2004; Law and Marles 2004; Lezaun 2004; Mohammed 2006; Morrow and Ingram 2005; Prudham 2007; Ziff 2005; Robertson 2005; Bernhardt 2005; Elliott 2006; Burrell and Hubicki 2005; Carolan 2008). To my knowledge, this chapter is the only effort to treat the case in relation to social movement strategy.

2. Re: Application of Abitibi Co. (1982), 62 CPR (2d) 81.

3. For a detailed overview of the intellectual property system for plants in Canada, see Benda 2003.

4. In the European Union, the patent system recognizes the concept of "ordre public," which refers to the public good and morality. Third parties are able to object to patent applications and issued patents on these grounds. Shobita Parthasarathy (2011), who has extensively studied the US and EU patent systems, observes that the US patent system has high "expertise barriers" compared to the European Union, making it difficult for outsiders to engage as equals in debates about patenting. Although there is not a parallel in-depth study of Canada's patenting system, activists in Canada clearly viewed the patent system as difficult to access. That said, the Canadian patent system seems to have lower barriers to access than the US one, as evidenced by the success of "life patent" opponents in the Harvard mouse case.

5. For a discussion of the NGO response to the CBAC, see Prudham and Morris 2006. For additional discussion of the low level of meaningful public participation in biotechnology policy in Canada, see Hartley and Skogstad 2005.

6. For a recent study of seed-saving practices in Canada, see Phillips 2009.

7. Others, seeing that this was politically unlikely, advocated the development of systems of compensation for the communities and countries that provide genetic resources (Kloppenburg [1988] 2005). With the 1992 CBD, the latter approach gained dominance. The CBD affirms the "sovereign rights of nations" over biological diversity and encourages the commercial (rather than free) exchange of genetic resources.

8. For more information about these campaigns, see the CBAN Web site at http://www.cban.ca/About/History. See also studies by Peter Andrée (2011), R. Steven

Turner (2001), Emily Eaton (2009), Andre Magnan (2007), and Lisa Mills (2002), and read histories of these struggles from the perspective of activists involved in them (Olson 2005; Sharratt 2001).

9. For analysis of the Harvard mouse case in Canada, see Prudham 2007; Furlanetto 2003; Swenarchuk and Canadian Environmental Law Association 2003.

10. Some observers have compared the farmers' seed saving to other "piracy" cases involving digital material such as software and media files, in which there is wide disagreement about whether it is fair for companies to go after ordinary people who share copyrighted digital material (Associated Press 2005). Noting parallels between plant breeding and software development, Kloppenburg (2010) has argued in favor of open-source-style provisions for plant genetic material, akin to systems that have emerged for the sharing and collective development of software.

11. Runyon has since appeared in the documentary films *Food, Inc.* and *The World according to Monsanto*. Other farmers and farm businesses have also been featured in documentaries and reports about Monsanto's efforts to protect its intellectual property rights. No other farmer, however, spent as much time in court against Monsanto as Schmeiser did, so no other case reveals as much about how the courts deal with the science of transgene flow in legal debates about patents and seed saving.

12. Canola Council of Canada, http://www.canolacouncil.org/buy_canola.aspx (accessed June 21, 2011).

13. Ibid.

14. Canola is harvested by cutting down (swathing) the plant and allowing it to dry in piles on the field, much like hay. It is then combined—picked up with machinery that separates the seeds from the rest of the plant.

15. The forum, organized by the Third World Network (2002), an international network of organizations working on issues affecting the Global South, aimed to explore "the root causes of the crises of poverty, ecology, governance and globalization."

16. This opinion was based, in part, on Friesen and Van Acker's own analysis of commercially available canola seeds (Friesen, Nelson, and Van Acker 2003). The highest percentage of Roundup Ready plants that Friesen found in a sample representing a field cultivated by Schmeiser was 67 percent, and the lowest was zero. In court, this was compared to the findings presented by Monsanto, which found levels ranging between 95 and 98 percent.

17. These points about the credibility of the evidence are all made in the trial brief for the defendant (*Monsanto v. Schmeiser* 2000, par. 55–107).

Chapter 6

1. Elisabeth Abergel (2007, 175), a policy scholar who has written about the Canadian regulatory system, notes that in Canada, "Exports of crops such as herbicide-tolerant canola have dramatically decreased while the demand for low-input agri-

culture has risen. Import bans, moratoria, strict labeling rules, public and political resistance to GE foods around the world are undermining the success of agricultural biotechnology." Gabriela Pechlaner and Gerardo Otero (2008, 351) argue that resistance to agricultural biotechnology, particularly in Mexico and other developing countries, "could modify, or even derail . . . the unfolding [global] food regime."

2. The value of the global organic market grew from US$15.9 billion in 1999 to US$54.9 billion in 2009 (Willer and Kilcher 2011).

3. Sociologist Michael Carolan (2008) has referred to Monsanto's contradictory positions on the "naturalness" of GE crops as "ontological gerrymandering."

4. There is a large and growing body of legal scholarship on the concept of liability as related to biotechnology (Burrell and Hubicki 2005; Cullet 2006; Glenn 2004; Khoury and Smyth 2005; Rodgers 2003; Smyth and Kershen 2006). The contentious international negotiations for "liability and redress" provisions under the Cartagena Protocol on Biosafety have been a focal point for discussion of this topic (Jungcurt and Schabus 2010; Masood 1996).

5. Geographer Emily Eaton comments on this controversy, and on the role of farmer and consumer activism in the battle over Roundup Ready wheat, in a forthcoming book.

6. In July 2001, 210 major Canadian agricultural, environmental, and citizens' advocacy groups sent a letter to Prime Minister Jean Chrétien asking him to prevent the introduction of GE wheat into Canadian agriculture. Notable signatories included the Canadian Wheat Board, the grain-marketing company that has sole responsibility for the sale and marketing of western Canadian wheat. Other key adherents were the National Farmers Union, Greenpeace Canada, and the Canadian Health Coalition (Phillipson 2001). Then, in March 2003, the Council of Canadians, a public interest NGO, made an eleven-city tour across three provinces in order to talk with farmers and activists about the need to reject Roundup Ready wheat. The organization described its strategy of incorporating farmers into the opposition to GE crops: "This tour marked an important shift in the Council of Canadians' campaign against GE because, for the first time, we have engaged with farmers alongside consumers in the fight to protect the integrity of our food supply" (Council of Canadians 2003). This strategy meant focusing on the market implications along with the food safety and environmental concerns that the Council of Canadians had previously raised in its antibiotech advocacy.

7. The struggle over wheat continues. In 2009, Monsanto relaunched its GE wheat research program, and grain industry groups in Canada, Australia, and the United States pledged their support for commercializing GE wheat seeds. In response, 223 farmer and consumer groups in 26 countries signed the "Definitive Global Rejection of GM Wheat" statement. For details and documents related to these statements, see the Canadian Biotechnology Action Network, http://www.cban.ca/Resources/Topics/GE-Crops-and-Foods-Not-on-the-Market/Wheat.

8. The Lairds became leading voices for organic agriculture in Saskatchewan, establishing the Back to the Farm Research Foundation in 1973, in order to promote organic farming.

9. Many contemporary organic farming practices—such as growing peas to build soil fertility—originated with the work of Albert Howard, a British agricultural researcher, and his wife, Gabrielle Howard. Their investigations of soil fertility and Indian composting systems in the early decades of the twentieth century became the basis of a widely adopted system of composting as well as a philosophy of organic farming (Gieryn 1999). In the early 1940s, Howard published two books—*An Agricultural Testament* and *The Soil and Health*—that criticized the use of synthetic fertilizers and outlined alternative methods of improving soil fertility. Organic ideas arrived in Canada in the 1950s, through literature distributed from Europe and the United States. Canada's first organic organization, the Canadian Organic Soil Organization, was founded in 1953, producing films such as *Understanding the Living Soil* and *A Sense of Humus* (Hill and MacRae 1992). Throughout the 1960s, ideas about organic farming were shared through these films and lectures by Spencer Cheshire, the founder of the Canadian Organic Soil Association (renamed the Land Fellowship).

10. The federation sets out a set of four organic principles:

Organic Agriculture should sustain and enhance the health of soil, plant, animal, human and planet as one and indivisible. . . . Organic Agriculture should be based on living ecological systems and cycles, work with them, emulate them and help sustain them. . . . Organic Agriculture should build on relationships that ensure fairness with regard to the common environment and life opportunities. . . . Organic Agriculture should be managed in a precautionary and responsible manner to protect the health and well-being of current and future generations and the environment. (International Federation of Organic Agriculture Movements 2009)

11. The first Canadian national standard on organic agriculture was published in 1999, but organic certification remained in the hands of private organizations such as the OCIA. Ten years later, the Organic Products Regulations came into force, making adherence to national organic standards mandatory for foods bearing the organic label and requiring certification bodies like the OCIA to be accredited by the CFIA. A key feature of the Organic Products Regulations is consistency with the US national organic standards, which facilitate international trade in organic products.

12. The ease of growing organic canola would become one of the most contested claims in the farmers' lawsuit against Monsanto and Bayer. The biotechnology companies argued that there was no class of farmers affected by GE contamination, because few farmers ever grew organic canola anyway. This debate reached beyond the courtroom, as observers of the case took sides. In interviews while the case was still going through the courts, critics of the organic farmers' position questioned the feasibility of growing organic canola, even in the absence of GE contamination. Many viewed canola as nearly impossible to grow without herbicides or fertilizers. Organic farmers and researchers, however, said that this view stemmed from an ignorance of organic farming techniques. It may not have been a simple crop to grow organically, but the farmers I interviewed believed the challenge was worth it because the price was good for organic canola.

13. At the international level, organic farmers have taken a similar stance. The International Federation of Organic Agriculture Movements (2002) argues that

organic farmers should not be held responsible for accidental contamination via gene flow, nor should they accept GE crop contamination. The federation opposes GE crops in all agriculture.

14. The potential for using nuisance law as a framework for a liability regime for biotechnology in the United Kingdom has been examined at length by Christopher Rodgers (2003). Because the legal systems of Canada and the United Kingdom have much in common, Rodgers's insights are relevant. Under the law of nuisance, the affected person may not complain simply of a loss of profitability; they "must establish a property right that has been infringed" (ibid., 379). This would require showing that significant damages have been done to the land or property itself. But, as Rodgers (ibid., 385–386) points out, there is a question as to:

whether "damage" is an objectively established factor, or whether it depends upon the subjective intent of the owner as to the intended composition of the property damaged. If the genetic makeup of Xs crop has been altered by cross pollination caused by wind drift of GM seed onto his land, surely this is "damage," in the sense that the physical composition of the crop is no longer that which he intends and wishes? It has been changed without his consent.

15. Along with Friesen, Van Acker examined samples of twenty-seven different seed lots, collected from farmers who had purchased the seeds. Through spray tests, the researchers found that fourteen of the twenty-seven unique seed lots had contamination levels above the cultivar purity guidelines for certified seed (Friesen et al. 2003). Organic farmers and antibiotech activists responded to these and other (Downey and Beckie n.d.) findings with alarm. Critics in Canada and the United States have warned that it is important to keep part of the seed supply GE free. The position is summed up in a report by the Union of Concerned Scientists, which states: "Seeds will be our only recourse if the prevailing belief in the safety of genetic engineering proves wrong. Heedlessly allowing the contamination of traditional plant varieties with genetically engineered sequences amounts to a huge wager on our ability to understand a complicated technology that manipulates life at the most elemental level" (Mellon 2004, 12). The findings of transgenic contamination in pedigreed seed have also raised serious concerns about whether inedible or unsafe transgenic crops—such as pharmaceutical crops and plants intended for industrial use only—could make their way into the seed supply as well.

16. This unpublished study, undertaken by researchers at the AAFC, found that three out of fourteen different conventional varieties of canola taken directly from certified seed growers contained herbicide-tolerance genes (Downey and Beckie n.d.). The researchers concluded that the transgenic contamination of certified seed most likely resulted from earlier generations of seed (breeder or foundation seed) that contained transgenes but were not tested. That is, the contamination of certified seed, at the levels detected, could not have resulted solely from gene flow during seed production by certified seed growers. If the breeder seeds used to produce certified seed are tested for transgenes, levels of transgene contamination can be kept below the 0.25 percent level. Nevertheless, this and another study (Friesen, Nelson, and Van Acker 2003) appears to call into question the acceptability of even that low level of contamination. Downey and Beckie (n.d., 11)

note that even a small percentage of herbicide-tolerant canola seeds can produce a large population of volunteer plants in the subsequent year, when a different crop will be grown on that field—effectively making them weeds that are resistant to at least one herbicide. Friesen, Nelson, and Van Acker (2003) made similar observations.

17. For a more detailed discussion of the ruling, see Garforth and Ainslie 2006.

18. In a press release after the Supreme Court decision, the farmers explained that they would not be taking individual legal action: "Class action legislation is an important tool for groups of citizens to gain access to justice. When denied the benefit of class action procedure, individuals are unfairly pitted against large multinational corporations that have enormous resources" (Organic Agriculture Protection Fund 2008).

19. There was also one evident contradiction in the ruling, again involving Ho's affidavit. Under the law as the judge interpreted it, economic losses alone are not a sufficient basis for a nuisance finding (*Hoffman et al. v. Monsanto Canada Inc. et al.* 2005b, par. 72). Therefore, organic farmers would need to demonstrate that transgenes are harmful to health or the environment in order to win under the nuisance law. In her ruling, the judge said that the plaintiffs made no "allegations that transgenic canola was harmful per se or made organic canola unfit for consumption" (Smyth and Kershen 2006, 13). But in fact, such claims had been made in the Ho affidavit, which the judge earlier ruled was irrelevant to this stage of the proceedings.

20. For a detailed discussion of these and other legal liability regimes with respect to agricultural biotechnology, see Smyth and Kershen 2006. In an interesting demonstration of the transnational nature of these struggles over GE, a representative from the organization of organic farmers in Saskatchewan was invited to Denmark in 2004 for hearings dealing with the creation of coexistence rules for GE, conventional, and organic agriculture. She was asked to share the experience of Canadian organic farmers with transgenic contamination. Soon after those hearings, the Danish government became the first in the world to enact liability rules for biotechnology, with an act of Parliament titled Act on the Growing, Etc. of Genetically Modified Crops (summarized in Smyth and Kershen 2006). A liability regime has also been created in Switzerland (Cullet 2006).

References

Abergel, Elisabeth A. 2007. Trade, Science, and Canada's Regulatory Framework for Determining the Environmental Safety of GE Crops. In *Genetically Engineered Crops: Interim Policies, Uncertain Legislation*, ed. Iain E. P. Taylor, 173–206. New York: Haworth Food and Agricultural Products Press.

Abergel, Elisabeth A., and Katherine Barrett. 2002. Putting the Cart before the Horse: A Review of Biotechnology Policy in Canada. *Journal of Canadian Studies; Revue d'Etudes Canadiennes* 37 (3): 135–161.

Adams, John S. 2009. Critics Sow Doubt as "Farmer Protection Act" Hearing Nears. *Great Falls Tribune*. http://www.organicconsumers.org/articles/article_17360.cfm (accessed June 21, 2011).

AFP y Notimex. 2005. Pide México oficialmente reconocer su cocina como patrimonio universal. http://www.jornada.unam.mx/2005/09/20/a03n1gas.php (accessed June 5, 2007).

Ag Department Uproots Science; Vilsack Seeks Out Politically Congenial Scientific Opinion. 2010. *Wall Street Journal*, December 27.http://online.wsj.com/article/SB10001424052748703581204576033611631362824.html (accessed August 25, 2011).

Agrawal, Arun. 2005. *Environmentality: Technologies ofGgovernment and the Making of Subjects*. Durham, NC: Duke University Press.

Agriculture and Agri-Food Canada. 2010. Certified Organic Production in Canada 2008. http://www4.agr.gc.ca/AAFC-AAC/display-afficher.do?id=1276033208187&lang=eng#an1 (accessed July 12, 2011).

Aguirre González, Rosa Luz. 2004. *La biotecnología agrícola en México: Efectos de la propiedad intelectual y la bioseguridad*. Mexico City: Universidad Autónoma Metropolitana.

AJAGI, CECCAM, CENAMI, CONTEC, Grupo ETC, GRAIN, UNOSJO, ORAB et al. 2005. Nuevas Evidencias de Contaminación Transgénica del Maíz Campesino. *Collective Press Release, Red en Defensa del Maíz*, December 8.

Alapekkala, Outi. 2011. Europe Paves Way for GM Crop Bans. *Guardian*. http://www.guardian.co.uk/environment/2011/jul/06/europe-gm-crop-bans (accessed August 25, 2011).

Alcántara, Cynthia Hewitt de, ed. 1994. *Economic Restructuring and Rural Subsistence in Mexico: Corn and the Crisis of the 1980s.* San Diego: Ejido Reform Research Project, Center for U.S.-Mexican Studies, University of California at San Diego.

Altieri, Miguel A. 2004. Linking Ecologists and Traditional Farmers in the Search for Sustainable Agriculture. *Frontiers in Ecology and the Environment* 2 (1): 35–42.

Altieri, Miguel A., and Peter Rosset. 1999a. Strengthening the Case for Why Biotechnology Will Not Help the Developing World: A Response to McGloughlin. *AgBioForum* 2 (3–4): 226–236.

Altieri, Miguel A., and Peter Rosset. 1999b. Ten Reasons Why Biotechnology Will Not Ensure Food Security, Protect the Environment, and Reduce Poverty in the Developing World. *AgBioForum* 2 (3–4): 155–162.

Alvarez-Buylla, Elena R. 2004. *Ecological and Biological Aspects of the Impacts of Transgenic Maize, including Agro-Biodiversity* (report prepared for the Secretariat of the Commission for Environmental Cooperation of North America). Montreal: Commission for Environmental Cooperation. http://www.cec.org/Page.asp?PageID=924&ContentID=2791 (accessed January 7, 2012).

Alvarez-Morales, Ariel. 1995. Implementation of Biosafety Regulations in a Developing Country: The Case of Mexico. *African Crop Science Journal* 3 (3): 309–314.

Alvarez-Morales, Ariel. 2000. Mexico: Ensuring Environmental Safety while Benefiting from Biotechnology. In *Agricultural Biotechnology and the Poor: Proceedings of an International Conference*, ed. G. J. Persley and Manuel Montecer Lantin, 90–96. Washington, DC: Consultative Group on International Agricultural Research.

Andow, David A., and Claudia Zwahlen. 2006. Assessing Environmental Risks of Transgenic Plants. *Ecology Letters* 9: 196–214.

Andrée, Peter. 2002. The Biopolitics of Genetically Modified Organisms in Canada. *Journal of Canadian Studies; Revue d'Etudes Canadiennes* 37: 162–191.

Andrée, Peter. 2007. *Genetically Modified Diplomacy: The Global Politics of Agricultural Biotechnology and the Environment.* Vancouver: University of British Columbia Press.

Andrée, Peter. 2011. Civil Society and the Political Economy of GMO Failures in Canada: A Neo-Gramscian Analysis. *Environmental Politics* 20 (2): 173–191.

Andrée, Peter, and Lucy Sharratt. 2004. *Genetically Modified Organisms and Precaution: Is the Canadian Government Implementing the Royal Society of Canada's Recommendations?* A Report on the Canadian Government's Response to the Royal Society of Canada's Expert Panel Report Entitled "Elements of Precaution: Recommendations for the Regulation of Food Biotechnology in Canada." Ottawa: Polaris Institute.

Antal, Edit. 2006. Lessons from NAFTA: The Role of the North American Commission for Environmental Cooperation in Conciliating Trade and Environment. *Michigan State Journal of International Law* 14: 167–189.

Antoniou, Michael, Mohamed Ezz El-Din Mostafa Habib, C. Vyvyan Howard, Richard C. Jennings, Carlo Leifert, Rubens Onofre Nodari, Claire Robinson, and John Fagan. 2011. Roundup and Birth Defects: Is the Public Being Kept in the Dark? http://earthopensource.org/index.php/reports/17-roundup-and-birth-defects-is-the-public-being-kept-in-the-dark (accessed January 6, 2012).

Armstrong, Elizabeth A., and Mary Bernstein. 2008. Culture, Power, and Institutions: A Multi-Institutional Politics Approach to Social Movements. *Sociological Theory* 26 (1): 74–99.

Associated Press. 2005. Saving Seed Is Latest Tech Piracy. *Wired*, January 14. http://www.wired.com/science/discoveries/news/2005/01/66282 (accessed August 8, 2011).

Australian Broadcasting Corporation. 2005. Canola GM Contamination Traced to Tas Trial. http://www.abc.net.au/rural/content/2005/s1496131.htm (accessed June 22, 2007).

Australian Broadcasting Corporation. 2006. Human Error "Probable Cause" of GM Canola Mix-up. http://www.abc.net.au/rural/news/content/2006/s1735224.htm (accessed July 1, 2007).

Babb, Sarah. 2005. The Social Consequences of Structural Adjustment: Recent Evidence and Current Debates. *Annual Review of Sociology* 31: 199–222.

Baker, Lauren E. 2008. Local Food Networks and Maize Agrodiversity Conservation: Two Case Studies from Mexico. *Local Environment* 13 (3): 235–251.

Ban Terminator. n.d. The Campaign. http://www.banterminator.org/The-Campaign (accessed July 13, 2011).

Ban Terminator Campaign. 2006. Terminator Seed Battle Begins: Farmers Face Billions of Dollars in Potential Costs. http://www.etcgroup.org/en/node/23 (accessed July 13, 2011).

Barcenas, Arturo Cruz. 2005. Fallo en contra de la comida mexicana como patrimonio de la humanidad. http://www.jornada.unam.mx/2005/11/26/a10n1gas.php (accessed June 5, 2007).

Barkan, Steven E. 1985. *Protesters on Trial: Criminal Justice in the Southern Civil Rights and Vietnam Antiwar Movements*. New Brunswick, NJ: Rutgers University Press.

Barkan, Steven E. 2006. Criminal Prosecution and the Legal Control of Protest. *Mobilization: An International Quarterly* 11 (2): 181–194.

Barkin, David. 1987. The End to Food Self-sufficiency in Mexico. *Latin American Perspectives* 14 (3): 271–297.

Barkin, David. 2002. The Reconstruction of a Modern Mexican Peasantry. *Journal of Peasant Studies* 30 (1): 73–90.

Barlett, Donald L., and James B. Steele. 2008. Monsanto's Harvest of Fear. *Vanity Fair*, May. http://www.vanityfair.com/politics/features/2008/05/monsanto200805?currentPage=1 (accessed August 8, 2011).

Barreda, Andrés. 2003. Biopiracy, Bioprospecting, and Resistance: Four Cases in Mexico. In *Contronting Globalization*, ed. Timothy A. Wise, Hilda Salazar, and Laura Carlsen, 101–128. Bloomfield, CT: Kumarian Press.

Barrett, Katherine, and Elisabeth A. Abergel. 2000. Breeding Familiarity: Environmental Risk Assessment for Genetically Engineered Crops in Canada. *Science and Public Policy* 27 (1): 2–12.

Barrett, Katherine, and Elisabeth A. Abergel. 2002. Defining a Safe Genetically Modified Organism: Boundaries of Scientific Risk Assessment. *Science and Public Policy* 29 (1): 47–58.

Barry, Tom. 1995. *Zapata's Revenge: Free Trade and the Farm Crisis in Mexico.* Boston: South End Press.

Bartra, Armando. 2003. Milpas of the Millennium: Where Will the Excluded Ones Go If the System Is Global? *Research in Rural Sociology and Development* 9: 35–42.

Beck, Ulrich. 1992. *Risk Society: Towards a New Modernity.* London: Sage Publications.

Beck, Ulrich. 1995. *Ecological Politics in an Age of Risk.* Cambridge: Polity.

Beck, Ulrich. 1999. *World Risk Society.* Cambridge: Polity.

Beckie, Hugh J., Suzanne I. Warwick, Harikumar Nair, and Ginette Seguin-Swartz. 2003. Gene Flow in Commercial Fields of Herbicide-Resistant Canola (Brassica Napus). *Ecological Applications* 13 (5): 1276–1294.

Bellon, Mauricio R., and Julien Berthaud. 2006. Traditional Mexican Agricultural Systems and the Potential Impacts of Transgenic Varieties on Maize Diversity. *Agriculture and Human Values* 23: 3–14.

Belson, Ken. 2011. Doubting Assurances, Japanese Find Radioactivity on Their Own. *New York Times*, August 1, A1.

Benda, Stan. 2003. The Sui Generis System for Plants in Canada: Quirks and Quarks of Seeds, Suckers, Splicing, and Brown-Bagging for the Novice. *Canadian Intellectual Property Review* 20: 323.

Benford, Robert D., and David A. Snow. 2000. Framing Processes and Social Movements: An Overview and Assessment. *Annual Review of Sociology* 26: 611–639.

Benveniste, Guy. 1973. *The Politics of Expertise.* Berkeley: Glendessary Press.

Bereano, Philip L. 1997. Reflections of a Participant-Observer: The Technocratic/Democratic Contradiction in the Practice of Technology Assessment. *Technological Forecasting and Social Change* 54 (2–3): 163–175.

Bereano, Philip L., and Martin Phillipson. 2004. Goliath v. Schmeiser. *Genewatch* 17 (4).

Bernauer, Thomas. 2003. *Genes, Trade, and Regulation: The Seeds of Conflict in Food Biotechnology.* Princeton, NJ: Princeton University Press.

Bernhardt, Stephanie M. 2005. High Plains Drifting: Wind-Blown Seeds and the Intellectual Property Implications of the GMO Revolution. *Northwestern Journal of Technology and Intellectual Property* 4 (1): 1–13.

Billings, Paul R., and Peter Shorett. 2007. Coping with Uncertainty: The Human Health Implications of GE Foods. In *Genetically Engineered Crops: Interim Poli-*

cies, Uncertain Legislation, ed. Iain E. P. Taylor, 75–90. New York: Haworth Food and Agricultural Products Press.

Bonanno, Alessandro, Lawrence Busch, William H. Friedland, Lourdes Gouveia, and Enzo Mingione, eds. 1994. *From Columbus to ConAgra: The Globalization of Agriculture and Food*. Lawrence: University Press of Kansas.

Bondera, Melanie, and Mark Query. 2006. Hawaiian Papaya: GMO Contaminated. http://www.hawaiiseed.org/downloads/publications-and-reports/Papaya-Contamination-Report.pdf/view (accessed August 25, 2011).

Boyd, William. 2003. Wonderful Potencies? Deep Structure and the Problem of Monopoly in Agricultural Biotechnology. In *Engineering Trouble: Biotechnology and Its Discontents*, ed. Rachel A. Schurman and Dennis D. Takahashi Kelso, 24–62. Berkeley: University of California Press.

Brand, Ulrich, and Christoph Gorg. 2003. The State and the Regulation of Biodiversity: International Biopolitics and the Case of Mexico. *Geoforum* 34: 221–233.

Brookey, Robert Alan. 2002. *Reinventing the Male Homosexual: The Rhetoric and Power of the Gay Gene*. Bloomington: Indiana University Press.

Brown, Jennifer. 2004. *Ejidos and Comunidades in Oaxaca, Mexico: Impact of the 1992 Reforms*. Reports on Foreign Aid and Development. Brandon, MB: Rural Development Institute.

Brown, Phil. 1987. Popular Epidemiology: Community Response to Toxic Waste-Induced Disease in Woburn, Massachusetts. *Science, Technology, and Human Values* 12 (3–4): 78–85.

Brown, Phil. 1992. Popular Epidemiology and Toxic Waste Contamination: Lay and Professional Ways of Knowing. *Journal of Health and Social Behavior* 33 (3): 267–281.

Brown, Phil. 2007. *Toxic Exposures: Contested Illnesses and the Environmental Health Movement*. New York: Columbia University Press.

Brown, Phil, Brian Mayer, Stephen Zavestoski, Theo Luebke, Joshua Mandelbaum, and Sabrina McCormick. 2003. The Health Politics of Asthma: Environmental Justice and Collective Illness Experience in the United States. *Social Science and Medicine* 57 (3): 453–464.

Brown, Phil, and Edwin J. Mikkelsen. 1990. *No Safe Place: Toxic Waste, Leukemia, and Community Action*. Berkeley: University of California Press.

Brunk, Conrad G. 2006. Public Knowledge, Public Trust: Understanding the "Knowledge Deficit." *Community Genetics* 9: 178–183.

Brush, Stephen, and Michelle Chauvet. 2004. Assessment of Social and Cultural Effects Associated with Transgenic Maize Production. In *Maize and Biodiversity: Background Volume*. Montreal: Commission for Environmental Cooperation. http://www.cec.org/Page.asp?PageID=924&ContentID=2796 (accessed January 6, 2012).

Bullard, Robert D., ed. 2005. *The Quest for Environmental Justice: Human Rights and the Politics of Pollution*. San Francisco: Sierra Club Books.

Burdge, Rabel J. 2003. The Practice of Social Impact Assessment-Background. *Impact Assessment and Project Appraisal* 21 (2): 84–88.

Burrell, Robert, and Stephen Hubicki. 2005. Patent Liability and Genetic Drift. *Environmental Law Review* 7 (4): 278–286.

Busch, Lawrence. 1991. *Plants, Power, and Profit: Social, Economic, and Ethical Consequences of the New Biotechnologies*. Cambridge: Basil Blackwell.

Busch, Lawrence, Valerie Gunter, Theodore Mentele, Masashi Tachikawa, and Keiko Tanaka. 1994. Socializing Nature: Technoscience and the Transformation of Rapeseed into Canola. *Crop Science* 34 (3): 607–614.

Busch, Lawrence, and Arunas Juska. 1997. Beyond Political Economy: Actor Networks and the Globalization of Agriculture. *Review of International Political Economy* 4 (4): 688–708.

Busch, Nathan A. 2002. Jack and the Beanstalk: Property Rights in Genetically Modified Plants. *Minnesota Intellectual Property Review* 3 (1): 1–235.

Buttel, Frederick H. 1995. The Global Impacts of Agricultural Biotechnology: A Post-Green Revolution Perspective. In *Issues in Agricultural Bioethics*, ed. T. B. Mepham, Gregory A. Tucker, and Julian Wiseman, 345–360. Nottingham: Nottingham University Press.

Buttel, Frederick H. 2005. The Environmental and Post-environmental Politics of Genetically Modified Crops and Foods. *Environmental Politics* 14 (3): 309–323.

Buttel, Frederick H., and Jill M. Belsky. 1987. Biotechnology, Plant Breeding, and Intellectual Property: Social and Ethical Dimensions. *Science, Technology, and Human Values* 12 (1): 31–49.

Buttel, Frederick H., J. Tadlock Cowan, Martin Kenney, and Jack Kloppenburg Jr. 1984. Biotechnology in Agriculture: The Political Economy of Agribusiness Reorganization and Industry-University Relationships. In *Research in Rural Sociology and Development*, ed. Harry Schwarzweller, 1: 315–348. Greenwich, CT: JAI Press.

Campbell, John L., and Ove K. Pedersen. 2001. Introduction: The Rise of Neoliberalism and Institutional Analysis. In *The Rise of Neoliberalism and Institutional Analysis*, ed. John L. Campbell and Ove K. Pedersen, 1–23. Princeton, NJ: Princeton University Press.

Canola Council of Canada. 2008. Canola Socioeconomic Value Report. http://www.canolacouncil.org/canadian_canola_industry.aspx (accessed June 8, 2010).

Čapek, Stella M. 2000. Reframing Endometriosis: From "Career Woman's Disease" to Environment/Body Connections. In *Illness and the Environment: A Reader in Contested Medicine*, ed. Steve Kroll-Smith, Phil Brown, and Valerie J. Gunter, 345–363. New York: New York University Press.

Carolan, Michael S. 2008. From Patent Law to Regulation: The Ontological Gerrymandering of Biotechnology. *Environmental Politics* 17 (5): 749–765.

Carruthers, David V. 1996. Indigenous Ecology and the Politics of Linkage in Mexican Social Movements. *Third World Quarterly* 17 (5): 1007–1028.

Carruthers, David V. 1997. Agroecology in Mexico: Linking Environmental and Indigenous Struggles. *Society & Natural Resources* 10 (3): 259–272.

Castle, Linda A., Gusui Wu, and David McElroy. 2006. Agricultural Input Traits: Past, Present, and Future. *Current Opinion in Biotechnology* 17 (2): 105–112.

Caudill, David S., and Lewis H. LaRue. 2006. *No Magic Wand: The Idealization of Science in Law*. Lanham, MD: Rowman and Littlefield.

Center for Food Safety. 2000. The Hidden Health Hazards of Genetically Engineered Food. *Food Safety Review*. http://www.centerforfoodsafety.org/campaign/genetically-engineered-food/crops/other-resources (accessed May 18, 2011).

Center for Food Safety. 2005. *Monsanto vs. U.S. Farmers*. Washington, DC: Center for Food Safety.

Center for Food Safety. 2006. GE Alfalfa Lawsuit: *Geertson Seed Farms, et al. v. Mike Johanns et al.*—Executive Summary. http://www.centerforfoodsafety.org/campaign/genetically-engineered-food/crops/legal-actions (accessed May 18, 2011).

Center for Food Safety. 2007. Monsanto vs. U.S. Farmers: November 2007 Update. http://www.centerforfoodsafety.org/pubs/Monsanto%20November%20 2007%20update.pdf (accessed May 31, 2010).

Center for Food Safety. 2011a. Letter to Secretary Vilsack. http://www.centerforfoodsafety.org/campaign/genetically-engineered-food/crops/policy-comments (accessed May 18, 2011).

Center for Food Safety. 2011b. Press Release: Farmers and Consumer Groups File Lawsuit Challenging Genetically Engineered Alfalfa Approval. http://www.centerforfoodsafety.org/2011/03/18/farmers-and-consumer-groups-file-lawsuit-challenging-genetically-engineered-alfalfa-approval (accessed May 18, 2011).

Chandler, Jennifer A. 2007. The Autonomy of Technology: Do Courts Control Technology or Do They Just Legitimize Its Social Acceptance? *Bulletin of Science, Technology, and Society* 27 (5): 339–348.

CIBIOGEM [Comisión Intersecretarial de Bioseguridad de los Organismos Genéticamente Modificados]. 2004. Comments by CIBIOGEM Technical Committee. In *Maize and Biodiversity: The Effects of Transgenic Maize in Mexico*. Commission for Environmental Cooperation. Montreal: Communications Department of the CEC Secretariat.

Clapp, Richard W. 2002. Popular Epidemiology in Three Contaminated Communities. *Annals of the American Academy of Political and Social Science* 584: 35.

Clark, E. Ann. 2004. So, Who Really Won the *Schmeiser* Decision. Paper presented to the National Farmers Union, June 10, Milverton, Ontario.

Cleveland, David A., Daniela Soleri, Flavio Aragón Cuevas, José Crossa, and Paul Gepts. 2005. Detecting (Trans)gene Flow to Landraces in Centers of Crop Origin: Lessons from the Case of Maize in Mexico. *Environmental Biosafety Research* 4: 197–208.

Cockrall-King, Jennifer. 2011. Bill C-474 Was Defeated in Canada and All I Got Was This Lousy Email from My MP. http://foodgirl.squarespace.com/

book/2011/2/11/bill-c-474-was-defeated-in-canada-and-all-i-got-was-this-lou.
html (accessed January 6, 2012).

Cohn, Avery, Jonathan Cook, Margarita Fernández, Rebecca Reider, and Corrina Steward, eds. 2006. *Agroecology and the Struggle for Food Sovereignty in the Americas.* London: International Institute for Environment and Development.

Collier, George A., and Jane F. Collier. 2005. The Zapatista Rebellion in the Context of Globalization. *Journal of Peasant Studies* 32 (3): 450–460.

Commission for Environmental Cooperation. 2011a. About the Commission. http://www.cec.org/Page.asp?PageID=924&SiteNodeID=310 (accessed May 22, 2011).

Commission for Environmental Cooperation. 2011b. Maize and Biodiversity. http://www.cec.org/Page.asp?PageID=924&SiteNodeID=347 (accessed May 22, 2011).

Commission of the European Communities. 2003. *Commission Recommendation of 23 July 2003 on Guidelines for the Development of National Strategies and Best Practices to Ensure the Co-existence of Genetically Modified Crops with Conventional and Organic Farming.* Brussels. http://ec.europa.eu/agriculture/publi/reports/coexistence2/index_en.htm (accessed January 6, 2012).

Conko, Gregory, and Henry I. Miller. 2010. The Environmental Impact Subterfuge. *Nature Biotechnology* 28 (12): 1256–1258.

Cooke, Bill, and Uma Kothari, eds. 2001. *Participation: The New Tyranny?* New York: Zed Books.

Corburn, Jason. 2005. *Street Science: Community Knowledge and Environmental Health Justice.* Cambridge: MIT Press.

Cornelius, Wayne A., and David Myhre, eds. 1998. *The Transformation of Rural Mexico: Reforming the Ejido Sector.* San Diego: University of California Press.

Couch, Stephen R., and Steve Kroll-Smith. 2000. Environmental Movements and Expert Knowledge: Evidence for a New Populism. In *Illness and the Environment: A Reader in Contested Medicine*, ed. Steve Kroll-Smith, 384–408. New York: New York University Press.

Council of Canadians. 2003. Planting Seeds of Doubt. Canadian Perspectives. Spring. http://www.canadians.org/publications/CP/2003/spring/seeds.html (accessed June 28, 2007).

CP Wire. 2000. Monsanto Hired Saskatchewan Prof to See How Far Its Seeds Can Fly. June 8. http://www.biotech-info.net/monsanto_schmeiser.html (accessed June 28, 2007).

CropLife Canada. 2010. Canada's Plant Science Industry Decries Second Reading Passage of Bill C-474. http://www.croplife.ca/web/english/mediaroom/newsreleases/2010/2010april15.cfm (accessed June 8, 2010).

Cullet, Philippe. 2006. Liability and Redress for Modern Biotechnology. *Yearbook of International Environmental Law* 15: 165–195.

Dalton, Rex. 2001. Transgenic Corn Found Growing in Mexico. *Nature* 413 (6854): 337.

de Beer, Jeremy. 2007. Rights and Responsibilities of Biotech Patent Owners. *University of British Columbia Law Review* 40: 343.

Delborne, Jason A. 2005. Pathways of Scientific Dissent in Agricultural Biotechnology. PhD diss., University of California at Berkeley.

Delborne, Jason A. 2008. Transgenes and Transgressions: Scientific Dissent as Heterogeneous Practice. *Social Studies of Science* 38 (4): 509–541.

DeSantis, S'ra. 2003. *Control through Contamination: US Forcing GMO Corn and Free Trade on Mexico and Central America*. Institute for Social Ecology Biotechnology Project and ACERCA. http://www.iatp.org/files/Control_Through_Contamination.pdf (accessed January 6, 2012).

Desmarais, Annette-Aurélie. 2002. The Vía Campesina: Consolidating an International Peasant and Farm Movement. *Journal of Peasant Studies* 29 (2): 91–124.

Dona, Artemis, and Ioannis S. Arvanitoyannis. 2009. Health Risks of Genetically Modified Foods. *Critical Reviews in Food Science and Nutrition* 49 (2): 164–175.

Douglas, Mary. (1966) 1978. *Purity and Danger: An Analysis of Concepts of Pollution and Taboo*. London: Routledge.

Downey, R. Keith. 2001. Scientific Societies Should Know Better. *National Post* (Canada), June 13, C19.

Downey, R. Keith, and Hugh J. Beckie. n.d. *Report on Project Entitled Isolation Effectiveness in Canola Pedigree Seed Production*. Saskatoon, SK: Agriculture and Agri-Food Canada.

Dreifus, Claudia. 2008. A Conversation with Nina V. Fedoroff, an Advocate for Science Diplomacy. *New York Times*. http://www.nytimes.com/2008/08/19/science/19conv.html (accessed August 25, 2011).

Drori, Gili S., and John W. Meyer. 2006. Global Scientization: An Environment for Expanded Organization. In *Globalization and Organization: World Society and Organizational Change*, ed. Gili S. Drori, John W. Meyer, and Hokyu Hwang, 50–68. Oxford: Oxford University Press.

Drori, Gili S., John W. Meyer, Francisco O. Ramirez, and Evan Schofer. 2003. *Science in the Modern World Polity: Institutionalization and Globalization*. Stanford, CA: Stanford University Press.

Dunlap, Riley E., and Angela G. Mertig. 1992. The Evolution of the U.S. Environmental Movement from 1970–1990: An Overview. *Society and Natural Resources* 4: 209–218.

Dyer, George A., J. Antonio Serratos-Hernández, Hugh R. Perales, Paul Gepts, Alma Piñeyro-Nelson, Angeles Chávez, Noé Salinas-Arreortua, Antonio Yúnez-Naude, J. Edward Taylor, and Elena R. Alvarez-Buylla. 2009. Dispersal of Transgenes through Maize Seed Systems in Mexico. *PLoS ONE* 4 (5): e5734.

Eaton, Emily. 2009. Getting behind the Grain: The Politics of Genetic Modification on the Canadian Prairies. *Antipode* 41 (2): 256–281.

Eaton, Emily. 2011. Let the Market Decide? Canadian Farmers Fight the Logic of Market Choice in GM Wheat. *ACME: An International E-Journal for Critical Geographies* 10 (1): 107–131.

Efe News Services. 2003. ONG encuentran maiz de variedad "Starlink" en campos mexicanos. October 9. www.lexisnexis.com (accessed January 7, 2012).

Elliott, Charlene D. 2006. Unlabelled: Law, Language, and Genetically Modified Foods in Canada. *Canadian Journal of Communication* 31 (1): 247–254.

Ellstrand, Norman C. 2003. *Dangerous Liaisons? When Cultivated Plants Mate with Their Wild Relatives.* Baltimore: Johns Hopkins University Press.

Endres, A. Bryan, and Thomas P. Redick. 2006. Agricultural Regulatory Update: EPA Seeks CAFO Rule Comment and States Preempt Establishment of GM-Free Zones. *Agricultural Management Committee Newsletter* 10 (2): 2–6.

Environment Canada. 2004. Canadian Comments on the CEC Secretariat's Article 13 Report. In *Maize and Biodiversity: The Effects of Transgenic Maize in Mexico.* Commission for Environmental Cooperation. Montreal: Communications Department of the CEC Secretariat.

Epstein, Steven. 1996. *Impure Science: AIDS, Activism, and the Politics of Knowledge.* Berkeley: University of California Press.

Escobar, Arturo. 1998. Whose Knowledge, Whose Nature? Biodiversity, Conservation, and the Political Ecology of Social Movements. *Journal of Political Economy* 5: 53–82.

Espeland, Wendy Nelson. 1998. *The Struggle for Water: Politics, Rationality, and Identity in the American Southwest.* Chicago: University of Chicago Press.

Esteva, Gustavo, and Catherine Marielle. 2003. *Sin Maíz No Hay País.* Mexico City: Museo Nacional de Culturas Populares.

ETC Group. 2003. Nine Mexican States Found to Be GM Contaminated. http://www.etcgroup.org/en/node/145 (accessed June 28, 2007).

ETC Group. 2004. Canadian Supreme Court Tramples Farmers' Rights—Affirms Corporate Monopoly on Higher Life Forms. http://www.etcgroup.org/en/node/106 (accessed June 11, 2010).

ETC Group. 2005. The Genetic Shell Game, or Now You See It! Now You Don't! http://www.etcgroup.org/upload/publication/50/01/etcmaizenrfinal.pdf (accessed June 28, 2007).

ETC Group. n.d. ETC Group: A Brief History. http://www.etcgroup.org/en/about/History_of_etcgroup_page (accessed May 21, 2011).

European Stance on GMOs Condemned by the WTO. 2006. *European Report,* October 3, 61460. Academic One File (accessed March 14, 1012).

Ewen, Stanley W. B., and Arpad Pusztai. 1999. Effect of Diets Containing Genetically Modified Potatoes Expressing Galanthus Nivalis Lectin on Rat Small Intestine. *Lancet* 354 (9187): 1353–1354.

Expert Panel of the Royal Society of Canada. 2001. *Elements of Precaution: Recommendations for the Regulation of Food Biotechnology in Canada.* Ottawa: Royal Society of Canada.

Eyerman, Ron, and Andrew Jamison. 1991. *Social Movements: A Cognitive Approach.* University Park: Pennsylvania State University Press.

Ezcurra, Exequiel, and Jorge Soberón Mainero. 2002. Evidence of Gene Flow from Transgenic Maize to Local Varieties in Mexico. In *LMOs and the Environment: Proceedings of an International Conference*, ed. Organization for Economic Cooperation and Development, 289–295. Paris: OECD.

Fairhead, James, and Melissa Leach. 2003. *Science, Society, and Power: Environmental Knowledge and Policy in West Africa and the Caribbean*. Cambridge: Cambridge University Press.

Farnham, Timothy J. 2007. *Saving Nature's Legacy: Origins of the Idea of Biological Diversity*. New Haven, CT: Yale University Press.

Fischer, Frank. 2000. *Citizens, Experts, and the Environment*. Durham, NC: Duke University Press.

Fitting, Elizabeth. 2006a. Importing Corn, Exporting Labor: The Neoliberal Corn Regime, GMOs, and the Erosion of Mexican Biodiversity. *Agriculture and Human Values* 23 (1): 15–26.

Fitting, Elizabeth. 2006b. The Political Uses of Culture: Maize Production and the GM Corn Debates in Mexico. *Focaal: European Journal of Anthropology* 48: 17–34.

Fitting, Elizabeth. 2011. *The Struggle for Maize: Campesinos, Workers, and Transgenic Corn in the Mexican Countryside*. Durham, NC: Duke University Press.

Fox, Jonathan. 1993. *The Politics of Food in Mexico: State Power and Social Mobilization*. Ithaca, NY: Cornell University Press.

Freudenburg, William R. 1986. Social Impact Assessment. *Annual Review of Sociology* 12: 451–478.

Frickel, Scott. 2004a. Building an Interdiscipline: Collective Action Framing and the Rise of Genetic Toxicology. *Social Problems* 51 (2): 269–287.

Frickel, Scott. 2004b. *Chemical Consequences: Environmental Mutagens, Scientist Activism, and the Rise of Genetic Toxicology*. New Brunswick, NJ: Rutgers University Press.

Frickel, Scott, Sahra Gibbon, Jeff Howard, Joanna Kempner, Gwen Ottinger, and David Hess. 2010. Undone Science: Charting Social Movement and Civil Society Challenges to Research Agenda Setting. *Science, Technology, and Human Values* 35 (4): 444–473.

Friedmann, Harriet. 1982. The Political Economy of Food: The Rise and Fall of the Postwar International Food Order. *American Journal of Sociology* 88: S248–S286.

Friesen, Lyle F., Alison G. Nelson, and Rene C. Van Acker. 2003. Evidence of Contamination of Pedigreed Canola (Brassica Napus) Seedlots in Western Canada with Genetically Engineered Herbicide Resistance Traits. *Agronomy Journal* 95 (5): 1342–1347.

Furlanetto, Angela. 2003. The Harvard Mouse Case: Developments in the Patentability of Life Forms. June. http://www.cba.org/CBA/newsletters/ip-2003/ip2.aspx (accessed June 28, 2007).

Garforth, Kathryn, and Paige Ainslie. 2006. When Worlds Collide: Biotechnology Meets Organic Farming in Hoffman v Monsanto. *Journal of Environmental Law* 18 (3): 459–477.

Garforth, Kathryn, Hari Subramaniam, Aseet Dalvi, and Barbara Cuber. 2004. Case Note: *Percy Schmeiser and Schmeier Enterprises Ltd. v. Monsanto Canada Inc. Review of European Community and International Environmental Law* 13 (3): 340–346.

Gieryn, Thomas F. 1999. *Cultural Boundaries of Science: Credibility on the Line.* Chicago: University of Chicago Press.

Gilbreth, Chris, and Gerardo Otero. 2001. Democratization in Mexico: The Zapatista Uprising and Civil Society. *Latin American Perspectives* 28 (4): 7–29.

Glenn, Jane Matthews. 2004. Footloose: Civil Responsibility for GMO Gene Wandering in Canada. *Washburn Law Journal* 43: 547–573.

Goldman, Michael. 2004. Imperial Science, Imperial Nature: Environmental Knowledge for the World (Bank). In *Earthly Politics: Local and Global in Environmental Governance*, ed. Sheila Jasanoff and Marybeth Long Martello, 55–80. Cambridge, MA: MIT Press.

Gómez Alarcón, Tonantzin. 2000. *Los OGTs Llegaron Ya: Los Organismos Geneticamente Transformados: un asunto ambiental, politico, social, etico y de salud.* Mexico City: Grupo de Estudios Ambientales A.C.

González, Roberto J. 2001. *Zapotec Science: Farming and Food in the Northern Sierra of Oaxaca.* Austin: University of Texas Press.

González, Roberto J. 2006. GM Maize as Cultural Crisis: Reactions among Zapotec Farmers to Transgenic Crops. Paper presented at the Annual Meeting of the Society for Social Studies of Science, Vancouver.

Goodman, David. 2000. Regulating Organic: A Victory of Sorts. *Agriculture and Human Values* 17: 211–213.

Goodman, David. 2003. The Brave New Worlds of Agricultural Technoscience: Changing Perspectives, Recurrent Themes, and New Research Directions in Agro-Food Studies. In *Engineering Trouble: Biotechnology and Its Discontents*, ed. Rachel A. Schurman and Dennis Doyle Takahashi Kelso, 218–238. Berkeley: University of California Press.

Goodman, David, Bernardo Sorj, and John Wilkinson. 1987. *From Farming to Biotechnology: A Theory of Agro-Industrial Development.* New York: Basil Blackwell.

Goodman, David, and Michael J. Watts, eds. 1997. *Globalising Food: Agrarian Questions and Global Restructuring.* New York: Routledge.

Greenpeace International. 1999. Press Release: Greenpeace Confirms USA Is Introducing Transgenic Maize into Mexico (May 25, 1999). http://78.47.137.204/genet/1999/May/msg00104.html (accessed September 24, 2008).

Greenpeace International. 2004. Monsanto Wins Right to Genetic Pollution. http://www.greenpeace.org/international/en/news/features/monsanto-wins-right-to-pollute (accessed June 11, 2010).

Greenpeace International. 2006. Biggest Russian Food and Feed Importers Adopt GE Free Policy. http://www.greenpeace.org/international/en/press/releases/biggest-russian-food-and-fee (accessed July 7, 2011).

Greenpeace International. 2007. Say No to Genetic Engineering. http://www.greenpeace.org/international/en/campaigns/agriculture/problem/genetic-engineering (accessed June 25, 2007).

Greenpeace International and GeneWatch UK. 2009. FP967 ("Triffid") Flax Has Been Grown Illegally in Canada and Exported around the Globe. http://www.gmcontaminationregister.org/index.php?content=nw_detail1 (accessed January 20, 2010).

Greenpeace International and GeneWatch UK. 2011. GM Contamination Register. http://gmcontaminationregister.org (accessed August 3, 2011).

Greenpeace México. 2001. *Boletín 0194: Científicos de todo el mundo llaman a tomar medidas para detener la contaminación genética del maíz mexicano.* November 29. Mexico City: Greenpeace Mexico.

Greenpeace México. 2005. Bolívar Zapata se burla de la Academia Mexicana de Ciencias. February 3. http://www.greenpeace.org/mexico/news/bol-var-zapata-se-burla-de-la (accessed July 1, 2007).

Greenpeace México. 2007. Sin maíz no hay país . . . ¡Pon a México en tu boca! http://www.greenpeace.org/mexico/es/Noticias/2007/Julio/sin-ma-z-no-hay-pa-s-pon (accessed January 6, 2012).

Greenpeace México. 2011. Logros de Greenpeace México. http://www.greenpeace.org/mexico/es/Quienes-somos/Logros-de--Greenpeace--Mexico (accessed May 21, 2011).

Greenpeace México, Grupo de Estudios Ambientales, and UNORCA. 2004. *Boletín 0419: Avalan Comunidades Oaxaqueñas y ONG la Recomendación del Comité Público de la CCA sobre la Contaminación del Maíz.* April 21. Mexico City: Greenpeace México.

Gruszczynski, Lukasz A. 2006. The Role of Science in Risk Regulation under the SPS Agreement. http://works.bepress.com/lukasz_gruszczynski/2 (accessed August 25, 2011).

Gupta, Aarti, and Robert Falkner. 2006. The Influence of the Cartagena Protocol on Biosafety: Comparing Mexico, China, and South Africa. *Global Environmental Politics* 6 (4): 23–55.

Guthman, Julie. 2004. *Agrarian Dreams: The Paradox of Organic Farming in California.* Berkeley: University of California Press.

Guthman, Julie. 2007. The Polanyian Way? Voluntary Food Labels as Neoliberal Governance. *Antipode* 39 (3): 456–478.

Gutiérrez González, Alicia. 2010. *The Protection of Maize under the Mexican Biosafety Law: Environment and Trade.* Göttingen: Universitätsverlag Göttingen.

Habermas, Jürgen. 1970. *Toward a Rational Society: Student Protest, Science, and Politics.* Boston: Beacon Press.

Hall, Linda. 2000. Pollen Flow between Herbicide-Resistant Brassica Napus Is the Cause of Multiple-Resistant B-Napus Volunteers. *Weed Science* 48 (6): 688–694.

Hartley, Matt. 2008. Grain Farmer Claims Moral Victory in Seed Battle Against Monsanto. *The Globe and Mail* (Canada). March 20. http://www.common dreams.org/archive/2008/03/20/7784 (accessed January 6, 2012).

Hartley, Sarah, and Grace Skogstad. 2005. Regulating Genetically Modified Crops and Foods in Canada and the United Kingdom: Democratizing Risk Regulation. *Canadian Public Administration* 48 (3): 305–327.

Harvey, David. 2005. *A Brief History of Neoliberalism.* New York: Oxford University Press.

Harvey, Neil. 1998. *The Chiapas Rebellion: The Struggle for Land and Democracy.* Durham, NC: Duke University Press.

Hayden, Cori. 2003. *When Nature Goes Public: The Making and Unmaking of Bioprospecting in Mexico.* Princeton, NJ: Princeton University Press.

Hays, Samuel P. 1987. *Beauty, Health, and Permanence: Environmental Politics in the United States, 1955–1985.* Cambridge: Cambridge University Press.

Heffernan, William. 2000. Concentration of Ownership and Control in Agriculture. In *Hungry for Profit: The Agribusiness Threat to Farmers, Food, and the Environment*, ed. Fred Magdoff, John B. Foster, and Frederick H. Buttel. New York: Monthly Review Press Books.

Heller, Chaia. 2001. From Risk to Globalization: Discursive Shifts in the French Debate about GMOs. *Medical Anthropology Quarterly* 15 (1): 25–28.

Heller, Chaia. 2004. Risky Science and Savoir Faire: Peasant Expertise in the French Debate over Genetically Modified Crops. In *The Politics of Food*, ed. Marianne Lien and Brigitte Nerlich, 81–99. New York: Berg.

Henriques, Gisele, and Raj Patel. 2003. *Policy Brief No. 7: Agricultural Trade Liberalization and Mexico.* Oakland, CA: Food First/Institute for Food and Development Policy.

Henriques, Gisele, and Raj Patel. 2004. *NAFTA, Corn, and Mexico's Agricultural Trade Liberalization.* Oakland, CA: Food First/Institute for Food and Development Policy.

Herbert, Martha R. 2005. Food Free of Genetic Engineering: More Than a Right. In *Rights and Liberties in the Biotech Age: Why We Need a Genetic Bill of Rights*, ed. Sheldon Krimsky and Peter Shorett, 57–70. Rowman and Littlefield.

Hess, David J. 1995. *Science and Technology in a Multicultural World: The Cultural Politics of Facts and Artifacts.* New York: Columbia University Press.

Hess, David J. 2007. *Alternative Pathways in Science and Technology: Activism, Innovation, and the Environment in an Era of Globalization.* Cambridge, MA: MIT Press.

Hess, David J. 2009. The Potentials and Limitations of Civil Society Research: Getting Undone Science Done. *Sociological Inquiry* 79 (3): 306–327.

Hess, David J., Steve Breyman, Nancy Campbell, and Brian Martin. 2008. Science, Technology, and Social Movements. In *The Handbook of Science and Tech-*

nology Studies, ed. Edward J. Hackett, Olga Amsterdamska, Michael Lynch, and Judy Wajcman, 473–498. 3rd ed. Cambridge, MA: MIT Press.

Higgins, Vaughan, and Geoffrey A. Lawrence. 2005. Introduction: Globalization and Agricultural Governance. In *Agricultural Governance: Globalization and the New Politics of Regulation*, ed. Vaughan Higgins and Geoffrey Lawrence. London: Routledge.

Hill, Stuart B., and Rod J. MacRae. 1992. Organic Farming in Canada. *Agriculture Ecosystems and Environment* 39 (1–2): 71–84.

Hoffman et al. v. Monsanto Canada Inc. et al. 2002. Affidavit of Dr. Rene Van Acker, Court of Queen's Bench, Judicial Centre of Saskatoon, Saskatchewan, Canada, Q.B. No. 67 of A.D. 2002.

Hoffman et al. v. Monsanto Canada Inc. et al. 2005a. Memorandum of Law on Behalf of the Prospective Appellants, Court of Appeal for Saskatchewan, No. 1148.

Hoffman et al. v. Monsanto Canada Inc. et al. 2005b. Ruling of Judge G. A. Smith, Queen's Bench for Saskatchewan, 2005 SKQB 225.

Hoffman et al. v. Monsanto Canada Inc. et al. 2006. Factum on Behalf of the Appellants, Court of Appeal for Saskatchewan, No. 1148.

Holbach, Martina, and Lindsay Keenan. 2005. *No Market for GM Labelled Food in Europe.* Greenpeace International. http://www.greenpeace.org/eu-unit/Global/eu-unit/reports-briefings/2009/3/no-market-for-gm-labelled-food.pdf (accessed January 6, 2012).

Holt-Gimenez, Eric. 2006. *Campesino a Campesino: Voices from Latin America's Farmer to Farmer Movement for Sustainable Agriculture.* Oakland, CA: Food First Books.

House Committee on Agriculture, US Congress. 2011. Press Release: Lucas, Chambliss, Roberts: USDA Sending Mixed Signals on Genetically Engineered Alfalfa. http://agriculture.house.gov/press/PRArticle.aspx?NewsID=1293 (accessed May 18, 2011).

Hubbard, Kristina, and Farmer to Farmer Campaign on Genetic Engineering. 2009. *Out of Hand: Farmers Face the Consequences of a Consolidated Seed Industry.* Stoughton, WI.

International Federation of Organic Agriculture Movements. 2002. Position on Genetic Engineering and Genetically Modified Organisms. http://www.ifoam.org/press/positions/ge-position.html (accessed June 9, 2010).

International Federation of Organic Agriculture Movements. 2009. Principles of Organic Agriculture. http://www.ifoam.org/about_ifoam/principles/index.html (accessed July 12, 2011).

International Work Group for Indigenous Affairs. 1997. *The Indigenous World, 1996–97.* Copenhagen: International Work Group for Indigenous Affairs.

Jackson, John P., Jr. 2001. *Social Scientists for Social Justice: Making the Case against Segregation.* New York: New York University Press.

James, Clive. 2009. ISAAA Brief 41—2009: Highlights—Global Status of Commercialized Biotech/GM Crops. http://www.isaaa.org/resources/publications/briefs/41/highlights/default.asp (accessed June 8, 2010).

James, Clive. 2010a. Biotech Facts and Trends: Mexico. http://www.isaaa.org/resources/publications/biotech_country_facts_and_trends/default.asp (accessed August 12, 2011).

James, Clive. 2010b. *Global Status of Commercialized Biotech/GM Crops: 2010—Executive Summary*. ISAAA brief no. 42. Ithaca, NY.

Jamison, Andrew. 1996. The Shaping of the Global Environmental Agenda: The Role of Non-Governmental Organisations. In *Risk, Environment, and Modernity: Towards a New Ecology*, ed. Scott Lash, Bronislaw Szerszynski, and Brian Wynne, 224–245. London: Sage Publications.

Jasanoff, Sheila. 1990. *The Fifth Branch: Science Advisers as Policymakers*. Cambridge, MA: Harvard University Press.

Jasanoff, Sheila. 1995. *Science at the Bar: Law, Science, and Technology in America*. Cambridge, MA: Harvard University Press.

Jasanoff, Sheila. 2003. Technologies of Humility: Citizen Participation in Governing Science. *Minerva* 41 (3): 223–244.

Jasanoff, Sheila. 2005. *Designs on Nature: Science and Democracy in Europe and the United States*. Princeton, NJ: Princeton University Press.

Jenkins, Craig, and Charles Perrow. 1977. Insurgency of the Powerless: Farm Worker Movements (1946–1972). *American Sociological Review* 42: 249–268.

Joint Public Advisory Committee. 2004. Re: Maize and Biodiversity Symposium of the Commission for Environmental Cooperation. http://www.cec.org/Page.asp?PageID=30101&ContentID=16038&SiteNodeID=282 (accessed January 6, 2012).

Jungcurt, Stefan, and Nicole Schabus. 2010. Liability and Redress in the Context of the Cartagena Protocol on Biosafety. *Review of European Community and International Environmental Law* 19 (2): 197–206.

Juska, Arunas, and Lawrence Busch. 1994. The Production of Knowledge and the Production of Commodities: The Case of Rapeseed Technoscience. *Rural Sociology* 59 (4): 581–597.

Kaiser, Jocelyn. 2008. Is the Drought Over for Pharming? *Science* 320 (5875): 473.

Keck, Margaret E., and Kathryn Sikkink. 1998. *Activists beyond Borders: Advocacy Networks in International Politics*. Ithaca, NY: Cornell University Press.

Kenney, Martin, and Frederick H. Buttel. 1985. Biotechnology: Prospects and Dilemmas for Third World Development. *Development and Change* 16 (1): 61–91.

Khagram, Sanjeev. 2004. *Dams and Development: Transnational Struggles for Water and Power*. Ithaca, NY: Cornell University Press.

Khoury, Lara, and Stuart J. Smyth. 2005. Reasonable Foreseeability and Liability in Relation to Genetically Modified Organisms. Paper presented at the Ninth

International Consortium of Applied Bioeconomy Research International Conference on Agricultural Biotechnology: Ten Years Later, July 6–10, Ravello, Italy.

Kinchy, Abby J. 2006. On the Borders of Post-War Ecology: Struggles over the Ecological Society of America's Preservation Committee, 1917–1946. *Science as Culture* 15: 23–44.

Kinchy, Abby J., and Daniel Lee Kleinman. 2003. Organizing Credibility: Discursive and Organizational Orthodoxy on the Borders of Ecology and Politics. *Social Studies of Science* 33 (6): 869–896.

Kinchy, Abby J., Daniel Lee Kleinman, and Robyn Autry. 2008. Against Free Markets, against Science? Regulating the Socioeconomic Effects of Biotechnology. *Rural Sociology* 73 (2): 147–179.

Kleinman, Daniel Lee, and Abby J. Kinchy. 2003. Boundaries in Science Policy Making: Bovine Growth Hormone in the European Union. *Sociological Quarterly* 44 (4): 577–595.

Kleinman, Daniel Lee, and Abby J. Kinchy. 2007. Against the Neoliberal Steamroller? The Biosafety Protocol and the Social Regulation of Agricultural Biotechnologies. *Agriculture and Human Values* 24 (2): 195–206.

Kleinman, Daniel Lee, Abby J. Kinchy, and Robyn Autry. 2009. Local Variation or Global Convergence in Agricultural Biotechnology Policy? A Comparative Analysis. *Science and Public Policy* 36 (5): 361–371.

Kloppenburg, Jack Ralph. (1988) 2005. *First the Seed: The Political Economy of Plant Biotechnology, 1492–2000.* Madison: University of Wisconsin Press.

Kloppenburg, Jack R. 2010. Impeding Dispossession, Enabling Repossession: Biological Open Source and the Recovery of Seed Sovereignty. *Journal of Agrarian Change* 10 (3): 367–388.

Kneen, Brewster. 1992. *The Rape of Canola: The Social Construction of Canola, 1950–1992.* Toronto: Dundurn Press Limited.

Kneen, Brewster. 2004. Court Confusion. *Ram's Horn* 221. http://ramshorn.ca/issue-221-june-2004 (accessed June 20, 2011).

Knuttila, Murray. 2003. Globalization, Economic Development, and Canadian Agricultural Policy. In *Farm Communities at the Crossroads: Challenge and Resistance*, ed. Harry P. Diaz, Joann Jaffe, and Robert Stirling, 289–302. Regina: Canadian Plains Research Center.

Kuyek, Devlin. 2002. *The Real Board of Directors: The Construction of Biotechnology Policy in Canada, 1980–2002.* Sorrento, BC: Ram's Horn.

Kuyek, Devlin. 2004. *Stolen Seeds: The Privatisation of Canada's Agricultural Biodiversity.* Sorrento, BC: Ram's Horn.

Kuyek, Devlin. 2007a. *Good Crop/Bad Crop: Seed Politics and the Future of Food in Canada.* Toronto: Between the Lines.

Kuyek, Devlin. 2007b. Sowing the Seeds of Corporate Agriculture: The Rise of Canada's Third Seed Regime. *Studies in Political Economy* 80: 31–54.

Law, Grace S., and Jennifer A. Marles. 2004. *Monsanto v. Schmeiser*: Patent Protection for Genetically Modified Genes and Cells in Canada. *Health Law Review* 13 (1): 44–47.

Leach, Melissa, and Ian Scoones. 2005. Science and Citizenship in a Global Context. In *Science and Citizens: Globalization and the Challenge of Engagement*, ed. Melissa Leach, Ian Scoones, and Brian Wynne, 15–38. London: Zed Books.

Levidow, Les. 1998. Democratizing Technology—or Technologizing Democracy? Regulating Agricultural Biotechnology in Europe. *Technology in Society* 20 (2): 211–226.

Levidow, Les. 1999. Britain's Biotechnology Controversy: Elusive Science, Contested Expertise. *New Genetics and Society* 18 (1): 47–64.

Levidow, Les. 2001. The GM Crops Debate: Utilitarian Bioethics? *Capitalism, Nature, Socialism* 12 (1): 44–55.

Levidow, Les. 2002. Ignorance-Based Risk Assessment? Scientific Controversy over GM Food Safety. *Science as Culture* 11 (1): 61–67.

Levidow, Les. 2003. Precautionary Risk Assessment of Bt Maize: What Uncertainties? *Journal of Invertebrate Pathology* 83 (2): 113–117.

Levidow, Les, and Susan Carr. 1997. How Biotechnology Regulation Sets a Risk/Ethics Boundary. *Agriculture and Human Values* 14 (1): 29–43.

Levidow, Les, Joseph Murphy, and Susan Carr. 2007. Recasting "Substantial Equivalence": Transatlantic Governance of GM Food. *Science, Technology, and Human Values* 32: 26–64.

Lezaun, Javier. 2004. Pollution and the Use of Patents: A Reading of *Monsanto v. Schmeiser*. In *Biotechnology: Between Commerce and Civil Society*, ed. Nico Stehr, 135–158. New Brunswick, NJ: Transaction Books.

Lezaun, Javier. 2006. Creating a New Object of Government: Making Genetically Modified Organisms Traceable. *Social Studies of Science* 36: 499–531.

Lind, David, and Elizabeth Barham. 2004. The Social Life of the Tortilla: Food, Cultural Politics, and Contested Commodification. *Agriculture and Human Values* 21 (1): 47–60.

Lockie, Stewart. 2001. SIA in Review: Setting the Agenda for Impact Assessment in the 21st Century. *Impact Assessment and Project Appraisal* 19 (4): 277–287.

Lynch, Michael, and Simon Cole. 2005. Science and Technology Studies on Trial: Dilemmas of Expertise. *Social Studies of Science* 35 (2): 269–311.

Maasen, Sabine, and Peter Weingart, eds. 2005. *Democratization of Expertise? Exploring Novel Forms of Scientific Advice in Political Decision-Making*. Dordrecht: Springer.

Magnan, Andre. 2007. Strange Bedfellows: Contentious Coalitions and the Politics of GM Wheat. *Canadian Review of Sociology and Anthropology/La Revue Canadienne de Sociologie et d'Anthropologie* 44 (3): 289–317.

Marielle, Catherine, and Lizy Peralta. 2007. *La Contaminación Transgénica del Maíz en México: Luchas Civiles en Defensa del Maíz y de la Soberanía Alimentaria*. Mexico City: Grupo de Estudios Ambientales.

Martinez Gómez, Francisco, and Robert Torres. 2001. Hegemony, Commodification, and the State: Mexico's Shifting Discourse on Agricultural Germplasm. *Agriculture and Human Values* 18 (3): 285–294.

Mascarenhas, Michael. 2007. Where the Waters Divide: First Nations, Tainted Water, and Environmental Justice in Canada. *Local Environment* 12 (6): 565–577.

Mascarenhas, Michael, and Lawrence Busch. 2006. Seeds of Change: Intellectual Property Rights, Genetically Modified Soybeans and Seed Saving in the United States. *Sociologia Ruralis* 46 (2): 122–138.

Masood, Ehsan. 1996. Liability Clause Blocks Talks on Biosafety Protocol. *Nature* 382 (6590): 384.

Massieu Trigo, Yolanda Cristina, and San Vicente Tello Adelita. 2006. El proceso de aprobación de la ley de bioseguridad: Política a la mexicana e interés nacional. *El Cotidiano* 21 (136): 39–51.

Mauro, Ian J., and Stéphane M. McLachlan. 2008. Farmer Knowledge and Risk Analysis: Postrelease Evaluation of Herbicide Tolerant Canola in Western Canada. *Risk Analysis* 28 (2): 463–476.

Mauro, Ian J., Stéphane M. McLachlan, and Rene C. Van Acker. 2009. Farmer Knowledge and A Priori Risk Analysis: Pre-release Evaluation of Genetically Modified Roundup Ready Wheat across the Canadian Prairies. *Environmental Science and Pollution Research* 16 (6): 689–701.

McAdam, Doug. 1982. *Political Process and the Development of Black Insurgency, 1930–1970*. Chicago: University of Chicago Press.

McAdam, Doug, Sidney Tarrow, and Charles Tilly. 2001. *Dynamics of Contention*. Cambridge: Cambridge University Press.

McAfee, Kathleen. 2003. Corn Culture and Dangerous DNA: Real and Imagined Consequences of Maize Transgene Flow in Oaxaca. *Journal of Latin American Geography* 2 (1): 18–42.

McAfee, Kathleen. 2008. Beyond Techno-Science: Transgenic Maize in the Fight over Mexico's Future. *Geoforum* 39 (1): 148–160.

McCann, Michael. 2006. Law and Social Movements: Contemporary Perspectives. *American Review of Law and Social Science* 2: 17–38.

McCarthy, John D., and Mayer N. Zald. 1977. Resource Mobilization and Social Movements: A Partial Theory. *American Journal of Sociology* 82: 1212–1241.

McCormick, Sabrina. 2006. The Brazilian Anti-Dam Movement: Knowledge Contestation as Communicative Action. *Organization and Environment* 19 (3): 321–346.

McCormick, Sabrina, Phil Brown, and Stephen Zavestoski. 2003. The Personal Is Scientific, the Scientific Is Political: The Public Paradigm of the Environmental Breast Cancer Movement. *Sociological Forum* 18 (4): 545–576.

McGloughlin, Martina. 1999. Ten Reasons Why Biotechnology Will Be Important to the Developing World. *AgBioForum* 2 (3–4): 163–174.

McLeod-Kilmurray, Heather. 2007. *Hoffman v. Monsanto*: Courts, Class Actions, and Perceptions of the Problem of GM Drift. *Bulletin of Science, Technology, and Society* 27 (3): 188–201.

McMichael, Philip, ed. 1995. *Food and Agrarian Orders in the World-Economy*. Westport, CT: Praeger.

McMichael, Philip. 2007. *Development and Social Change: A Global Perspective*. Thousand Oaks, CA: Pine Forge Press.

Mellon, Margaret. 2004. Seeds of Doubt. *Catalyst: The Magazine of the Union of Concerned Scientists* 3 (1): 5–6, 12.

Melucci, Alberto. 1996. *Challenging Codes: Collective Action in the Information Age*. Cambridge: Cambridge University Press.

Mercer, Kristin L., and Joel D. Wainwright. 2008. Gene Flow from Transgenic Maize to Landraces in Mexico: An Analysis. *Agriculture Ecosystems and Environment* 123 (1–3): 109–115.

Mezzalama, Monica, Jonathan H. Crouch, and Rodomiro Ortiz. 2010. Monitoring the Threat of Unintentional Transgene Flow into Maize Gene Banks and Breeding Materials. *Electronic Journal of Biotechnology* 13 (2): 6–7.

Miller, Clark A. 2001. Challenges in the Application of Science to Global Affairs: Contingency, Trust, and Moral Order. In *Changing the Atmosphere: Expert Knowledge and Environmental Governance*, ed. Clark A. Miller and Paul N. Edwards, 247–285. Cambridge, MA: MIT Press.

Mills, Lisa N. 2002. *Science and Social Context: The Regulation of Recombinant Bovine Growth Hormone in North America*. Montreal: McGill-Queen's University Press.

Millstone, Erik, Eric Brunner, and Sue Mayer. 1999. Beyond "Substantial Equivalence." *Nature* 401 (6753): 525–526.

Mohammed, Emir Aly Crowne. 2006. Rounding Up Plant Patents and Other Growing Patent Concerns: The Canadian Teachings of *Monsanto v. Schmeiser*. Social Science Research Network, July 11.

Monsanto. 2004. Monsanto to Realign Research Portfolio, Development of Roundup Ready Wheat Deferred. May 10. http://monsanto.mediaroom.com/index.php?s=43&item=241 (accessed June 28, 2007).

Monsanto. 2011a. Dave Runyon. http://www.monsanto.com/newsviews/Pages/dave-runyon.aspx (accessed June 21, 2011).

Monsanto. 2011b. Monsanto's Commitment: Farmers and Patents. http://www.monsanto.com/newsviews/Pages/commitment-farmers-patents.aspx (accessed June 16, 2011).

Monsanto v. Schmeiser. 2000. Trial Brief on Behalf of the Defendants, Percy Schmeiser and Schmeiser Enterprises Ltd., Federal Court, Trial Division, Court File No. T-1593–98.

Monsanto v. Schmeiser. 2001. Reasons for Judgment, Judge W. Andrew MacCay, FCT 256.

Monsanto v. Schmeiser. 2004. Supreme Court of Canada, Judgment of 21 May, SCC 34.

Mooney, Pat Roy. 1979. *Seeds of the Earth: A Private or Public Resource?* Ottawa: Inter Pares.

Moore, Kelly. 1996. Organizing Integrity: American Science and the Creation of Public Interest Organizations, 1955–1975. *American Journal of Sociology* 101 (6): 1592–1627.

Moore, Kelly. 2008. *Disrupting Science: Social Movements, American Scientists, and the Politics of the Military, 1945–1975.* Princeton, NJ: Princeton University Press.

Moore, Kelly, Daniel Lee Kleinman, David Hess, and Scott Frickel. 2010. Science and Neoliberal Globalization: A Political Sociological Approach. *Theory and Society* 40 (5): 505–532.

Morello-Frosch, Rachel, Stephen Zavestoski, Phil Brown, Rebecca Gasior Altman, Sabrina McCormick, and Brian Mayer. 2006. Embodied Health Movements: Responses to a "Scientized" World. In *The New Political Sociology of Science: Institutions, Networks, and Power,* ed. Scott Frickel and Kelly Moore, 244–271. Madison: University of Wisconsin Press.

Morrow, A. David, and Colin B. Ingram. 2005. Of Transgenic Mice and Roundup Ready Canola: The Decisions of the Supreme Court of Canada in *Harvard College v. Canada* and *Monsanto v. Schmeiser. UBC Law Review* 38 (1): 189–222.

Morton, Roger. 2005. Mexican Government Studies on Transgenes in Mexican Maize? *AgBioView.* http://www.agbioworld.org/newsletter_wm/index.php?caseid=archive&newsid=2435 (accessed January 6, 2012).

Müller, Birgit. 2006. Infringing and Trespassing Plants: Patented Seeds at Dispute in Canada's Courts. *Focaal: European Journal of Anthropology* 48: 83–98.

Murphy, Joseph, and Les Levidow. 2006. *Governing the Transatlantic Conflict over Agricultural Biotechnology: Contending Coalitions, Trade Liberalisation, and Standard Setting.* New York: Routledge.

Nadal, Alejandro. 2002. Corn in NAFTA: Eight Years After—A Research Report Prepared for the North American Commission for Environmental Cooperation. http://www.cec.org/files/pdf/ECONOMY/Corn-NAFTA_en.pdf (accessed June 26, 2007).

Nadal, Alejandro. 2003. Corn in NAFTA Eight Years After: Effects on Mexican Biodiversity. In *Greening NAFTA: The North American Commission for Environmental Cooperation,* ed. David L. Markell and John H. Knox, 152–172. Stanford, CA: Stanford University Press.

Nadal, Alejandro. 2006. Ventana de oportunidad. http://www.jornada.unam.mx/2006/10/18/index.php?section=opinion&article=033a1eco (accessed January 6, 2012).

National Farmers Union. 2000. National Farmers Union Policy on Genetically Modified (GM) Foods. http://www.nfu.ca/policy/GM_FOOD_POLICY.misc.pdf (accessed June 26, 2007).

National Research Council. 2004. *Safety of Genetically Engineered Foods: Approaches to Assessing Unintended Health Effects.* Washington, DC: National Research Council.

Nickel, Rod. 2010. Canada Moves to Revive Flax Exports after GMO Flap. January 8. http://uk.reuters.com/article/2010/01/08/flax-canada-gmo-idUKN0824305620100108 (accessed January 6, 2012).

NGO/CSO Forum for Food Sovereignty. 2002. Food Sovereignty: A Right for All. http://www.foodsovereignty.org/Resources/Archive/Forum.aspx (accessed January 6, 2012).

O'Brien, Amanda. 2011. Farmer Sues over GM Taint. *The Australian,* July 28. http://www.theaustralian.com.au/news/nation/farmer-sues-over-gm-taint/story-e6frg6nf-1226103022356 (accessed January 6, 2012).

O'Rourke, Dara, and Gregg P. Macey. 2003. Community Environmental Policing: Assessing New Strategies of Public Participation in Environmental Regulation. *Journal of Policy Analysis and Management* 22 (3): 383–414.

Ochoa, Enrique C. 2000. *Feeding Mexico: The Political Uses of Food since 1910.* Wilmington, DE: SR Books.

Offe, Claus. 1985. New Social Movements: Challenging the Boundaries of Institutional Politics. *Sociological Research* 52: 817–868.

Oliver, Pamela E. 2008. Repression and Crime Control: Why Social Movement Scholars Should Pay Attention to Mass Incarceration as a Form of Repression. *Mobilization: An International Quarterly* 13 (1): 1–24.

Olson, R. Dennis. 2005. Hard Red Spring Wheat at a Genetic Crossroad: Rural Prosperity or Corporate Hegemony? In *Controversies in Science and Technology: From Maize to Menopause,* ed. Daniel Lee Kleinman, Abby J. Kinchy and Jo Handelsman, 150–168. Madison: University of Wisconsin Press.

Organic Agriculture Protection Fund. 2002. Organic Farmers Gain Key Piece of Evidence in Class Action. http://oapf.saskorganic.com/news.html (accessed January 6, 2012).

Organic Agriculture Protection Fund. 2008. Individual Action Not the Way to Go. http://oapf.saskorganic.com (accessed July 13, 2011).

Ortiz-García, Sol, Exequiel Ezcurra, B. Schoel, Francisca Acevedo, Jorge Soberón Mainero, and Allison A. Snow. 2005. Absence of Detectable Transgenes in Local Landraces of Maize in Oaxaca, Mexico (2003–2004). *Proceedings of the National Academy of Sciences of the United States of America* 102: 12338–12343.

Ottinger, Gwen. 2010. Buckets of Resistance: Standards and the Effectiveness of Citizen Science. *Science, Technology, and Human Values* 35 (2): 244–270.

Overdevest, Christine, and Brian Mayer. 2008. Harnessing the Power of Information through Community Monitoring: Insights from Social Science. *Texas Law Review* 86 (7): 1493–1526.

Paarlberg, Robert L. 2002. The Real Threat to GM Crops in Poor Countries: Consumer and Policy Resistance to GM Foods in Rich Countries. *Food Policy* 27 (3): 247–250.

Paarlberg, Robert L. 2008. *Starved for Science: How Biotechnology Is Being Kept Out of Africa*. Cambridge, MA: Harvard University Press.

Parthasarathy, Shobita. 2010. Breaking the Expertise Barrier: Understanding Activist Strategies in Science and Technology Policy Domains. *Science and Public Policy* 37 (5): 355–367.

Parthasarathy, Shobita. 2011. Whose Knowledge? What Values? The Comparative Politics of Patenting Life Forms in the United States and Europe. *Policy Sciences* 44 (3): 267–288.

Pearson, Thomas. 2009. On the Trail of Living Modified Organisms: Environmentalism within and against Neoliberal Order. *Cultural Anthropology* 24 (4): 712–745.

Pechlaner, Gabriela, and Gerardo Otero. 2008. The Third Food Regime: Neoliberal Globalism and Agricultural Biotechnology in North America. *Sociologia Ruralis* 48 (4): 351–371.

Pelletier, David L. 2005. Science, Law, and Politics in FDA's Genetically Engineered Foods Policy: Scientific Concerns and Uncertainties. *Nutrition Reviews* 63 (6): 210–223.

Pelletier, David L. 2006. FDA's Regulation of Genetically Engineered Foods: Scientific, Legal, and Political Dimensions. *Food Policy* 31 (6): 570–591.

Perez, U. Matilde. 2003. Transgénico prohibido para consumo humano contaminó maíz en 9 estados. *La Jornada*, October 9. http://www.jornada.unam.mx/2003/10/09/044n1soc.php?origen=index.html&fly=2 (accessed January 6, 2012).

Phillips, Catherine. 2009. Canada's Evolving Seed Regime: Relations of Industry, State, and Seed Savers. *Environments: A Journal of Interdisciplinary Studies* 36 (1): 5–18.

Phillips, Peter W. B. 2003. The Economic Impact of Herbicide Tolerant Canola in Canada. In *The Economic and Environmental Impacts of Agbiotech: A Global Perspective*, ed. Nicholas G. Kalaitzandonakes, 119–139. New York: Kluwer Academic/Plenum Publishers.

Phillips, Peter W. B., and Grant E. Isaac. 2001. Regulating International Trade in Knowledge-Based Products. In *The Biotechnology Revolution in Global Agriculture: Invention, Innovation, and Investment in the Canola Sector*, ed. Peter W. B. Phillips and George G. Khachatourians. Wallingford, UK: CABI Publishing.

Phillips, Peter W. B., and Stuart J. Smyth. 2004. Managing the Value of New-Trait Varieties in the Canola Supply Chain in Canada. *Supply Chain Management: An International Journal* 9 (3–4): 313–322.

Phillipson, Martin. 2001. Agricultural Law: Containing the GM Revolution. *Biotechnology and Development Monitor* 48: 2–5. http://www.biotech-monitor.nl/4802.htm (accessed January 6, 2012).

Piñeyro-Nelson, Alma, Joost Van Heerwaarden, Hugo R. Perales, J. Antonio Serratos-Hernández, A. Rangel, M. B. Hufford, Paul Gepts, A. Garay Arroyo, R. Rivera Bustamante, and Elena R. Álvarez-Buylla. 2009a. Resolution of the Mexican

Transgene Detection Controversy: Error Sources and Scientific Practice in Commercial and Ecological Contexts. *Molecular Ecology* 18 (20): 4145–4150.

Piñeyro-Nelson, Alma, Joost Van Heerwaarden, Hugo R. Perales, J. Antonio Serratos-Hernández, A. Rangel, M. B. Hufford, Paul Gepts, A. Garay Arroyo, R. Rivera Bustamante, and Elena R. Álvarez-Buylla. 2009b. Transgenes in Mexican Maize: Molecular Evidence and Methodological Considerations for GMO Detection in Landrace Populations. *Molecular Ecology* 18 (4): 750–761.

Poitras, Manuel. 2008a. Social Movements and Techno-Democracy: Reclaiming the Genetic Commons. In *Food for the Few: Neoliberal Globalism and Biotechnology in Latin America*, ed. Gerardo Otero, 267–287. Austin: University of Texas Press.

Poitras, Manuel. 2008b. Unnatural Growth: The Political Economy of Biotechnology in Mexico. In *Food for the Few: Neoliberal Globalism and Biotechnology in Latin America*, ed. Gerardo Otero, 115–133. Austin: University of Texas Press.

Prakash, Aseem, and Kelly Kollman. 2003. Biopolitics in the EU and the US: A Race to the Bottom or Convergence to the Top? *International Studies Quarterly* 47 (4): 617–641.

Prakash, C. S. 2005. Duh . . . No GM Genes in Mexican Corn. *AgBioView*, August 9. http://www.agbioworld.org/newsletter_wm/index.php?caseid=archive&newsid =2398 (accessed January 6, 2012).

Preibisch, Kerry L., Gladys Rivera Herrejon, and Steve L. Wiggins. 2002. Defending Food Security in a Free-Market Economy: The Gendered Dimensions of Restructuring in Rural Mexico. *Human Organization* 61 (1): 68–79.

Price, Don K. 1965. *The Scientific Estate*. Cambridge, MA: Harvard University Press.

Provincial Court of Saskatchewan. 2005. *Schmeiser v. Monsanto* SC 18/04.

Prudham, Scott. 2007. The Fictions of Autonomous Invention: Accumulation by Dispossession, Commodification and Life Patents in Canada. *Antipode* 39 (3): 406–429.

Prudham, Scott, and Angela Morris. 2006. "Making the Market Safe" for GM Foods: The Case of the Canadian Biotechnology Advisory Committee. *Studies in Political Economy* 78: 145–175.

Public Patent Foundation. 2011. *First Amended Complaint: Organic Seed Growers and Trade Association, et al. v. Monsanto Company and Monsanto Technology*. http://www.pubpat.org/monsanto-seed-patents.htm (accessed January 6, 2012).

Puricelli, Sonia. 2010. *El Movimiento el Campo No Aguanta Más: Auge, Contradicciones y Declive (México 2002—2004)*. Mexico: Plaza y Valdés.

Qualman, Darrin. 2001. *The Farm Crisis and Corporate Power*. Ottawa: Canadian Centre for Policy Alternatives.

Qualman, Darrin, and Nettie Wiebe. 2002. *The Structural Adjustment of Canadian Agriculture*. Ottawa: Canadian Centre for Policy Alternatives.

Quist, David, and Ignacio H. Chapela. 2001. Transgenic DNA Introgressed into Traditional Maize Landraces in Oaxaca, Mexico. *Nature* 414 (6863): 541–543.

Raustiala, Kal, and David G. Victor. 2004. The Regime Complex for Plant Genetic Resources. *International Organization* 58 (2): 277–309.

Red en Defensa del Maíz. 2009. No to Transgenic Maize! (English Version). http://endefensadelmaiz.org/No-to-transgenic-maize.html (accessed January 6, 2012).

Redlin, David. 2003. Rationale for U.S. Changes to the Terms of Reference. Comments on the Terms of Reference and Provisional Outline for the CEC Secretariat's Article 13 Report *Maize and Biodiversity: The Effects of Transgenic Maize in Mexico.* http://www.cec.org/Page.asp?PageID=924&ContentID=2791 (accessed January 6, 2012).

Reuters. 2003. La batalla en México vs. maíz transgénico. *El Universal*, October 23, 2-B.

Ribeiro, Silvia. 2009. Monitoreo Infecto. La Jornada, August 1. http://www.jornada.unam.mx/2009/08/01/index.php?section=opinion&article=025a1eco (accessed June 3, 2011).

Rieger, Mary A., Michael Lamond, Christopher Preston, Stephen B. Powles, and Richard T. Roush. 2002. Pollen-Mediated Movement of Herbicide Resistance between Commercial Canola Fields. *Science* 296 (5577): 2386–2388.

Right Livelihood Award Foundation. 2007. Percy and Louise Schmeiser. http://www.rightlivelihood.org/schmeiser.html (accessed June 11, 2010).

Robertson, Sean. 2005. Re-Imagining Economic Alterity: A Feminist Critique of the Juridical Expansion of Bioproperty in the Monsanto Decision at the Supreme Court of Canada. *University of Ottawa Law and Technology Journal* 2 (2): 227–253.

Rodgers, Christopher P. 2003. Liability for the Release of GMOs into the Environment: Exploring the Boundaries of Nuisance. *Cambridge Law Journal* 62 (2): 371–402.

Rosenburg, Gerald N. 1991. *The Hollow Hope: Can Courts Bring about Social Change?* Chicago: University of Chicago Press.

Sanchez Albarrán, Armando. 2004. Del movimiento ¡El campo no aguanta más! a las movilizaciones sociales en la cumbre de la OMC en Cancún. Dependencia o soberanía alimentaria: ésa es la cuestión . . . agraria. *El Cotidiano* 124.

Sarewitz, Daniel. 2004. How Science Makes Environmental Controversies Worse. *Environmental Science and Policy* 7 (5): 385–403.

Saskatchewan Organic Directorate. 2006. *Recent Activities and Achievements Report.* http://saskorganic.com/article/recent-activities-and-achievements-report (accessed January 6, 2012).

Scheingold, Stuart A. 1974. *The Politics of Rights: Lawyers, Public Policy, and Political Change.* New Haven, CT: Yale University Press.

Schnaiberg, Allan. 1980. *The Environment: From Surplus to Scarcity.* Oxford: Oxford University Press.

Schurman, Rachel A. 2003. Introduction: Biotechnology in the New Millennium. In *Engineering Trouble: Biotechnology and Its Discontents*, ed. Rachel A. Schurman and Dennis Doyle Takahashi Kelso, 1–23. Berkeley: University of California Press.

Schurman, Rachel. 2004. Fighting "Frankenfoods": Industry Opportunity Structures and the Efficacy of the Anti-Biotech Movement in Western Europe. *Social Problems* 51 (2): 243–268.

Schurman, Rachel A., and William Munro. 2006. Ideas, Thinkers, and Social Networks: The Process of Grievance Construction in the Anti-Genetic Engineering Movement. *Theory and Society* 35 (1): 1–38.

Schurman, Rachel A., and William A. Munro. 2010. *Fighting for the Future of Food: Activists versus Agribusiness in the Struggle over Biotechnology*. Minneapolis: University of Minnesota Press.

Secretariat of the Commission for Environmental Cooperation [CEC]. 2004. *Maize and Biodiversity: The Effects of Transgenic Maize in Mexico*. Montreal: Communications Department of the CEC Secretariat.

Seidman, Gay W. 2007. *Beyond the Boycott: Labor Rights, Human Rights, and Transnational Activism*. New York: Russell Sage Foundation.

Serratos-Hernández, José-Antonio, José-Luis Gómez-Olivares, Noé Salinas-Arreortua, Enrique Buendía-Rodríguez, Fabián Islas-Gutiérrez, and Ana de-Ita. 2007. Transgenic Proteins in Maize in the Soil Conservation Area of Federal District, México. *Frontiers in Ecology and the Environment* 5 (5): 247–252.

Shane, Kristen. 2010. Biotechnology Agriculture Industry Lobbies Hard against Atamanenko's Private Member's Bill. http://gefreebc.wordpress.com/2010/10/01/bill-c-474-alex-atamanenko-cban (accessed July 10, 2011).

Sharratt, Lucy. 2001. No to Bovine Growth Hormone: Ten Years of Resistance in Canada. In *Redesigning Life?: The Worldwide Challenge to Genetic Engineering*, ed. Brian Tokar. London: Zed Books.

Shiva, Vandana. 1995. Biotechnological Development and the Conservation of Biodiversity. In *Biopolitics: A Feminist and Ecological Reader on Biotechnology*, ed. Vandana Shiva and Ingunn Moser, 193–213. London: Zed Books.

Smits, Martijntje. 2006. Taming Monsters: The Cultural Domestication of New Technology. *Technology in Society* 28: 489–504.

Smyth, Stuart J., and Drew L. Kershen. 2006. Agricultural Biotechnology: Legal Liability Regimes from Comparative and International Perspectives. *Global Jurist Advances* 6 (2): 1–78.

Snow, Allison. 2009. Unwanted Transgenes Re-discovered in Oaxacan Maize. *Molecular Ecology* 18 (4): 569–571.

Snow, Allison A., David A. Andow, Paul Gepts, Eric M. Hallerman, A. Power, James M. Tiedje, and L. LaReesa Wolfenbarger. 2005. Genetically Engineered Organisms and the Environment: Current Status and Recommendations. *Ecological Applications* 15 (2): 377–404.

Soleri, Daniela, David A. Cleveland, and Flavio Aragón Cuevas. 2006. Transgenic Crops and Crop Varietal Diversity: The Case of Maize in Mexico. *Bioscience* 56 (6): 503–513.

Stabinsky, Doreen. 2000. Bringing Social Analysis into a Multilateral Environmental Agreement: Social Impact Assessment and the Biosafety Protocol. *Journal of Environment and Development* 9 (3): 260.

Stokstad, Erik. 2002. A Little Pollen Goes a Long Way. *Science* 296, June 28, 2314.

Straka, Allison M. 2010. Geertson Seed Farms v. Johanns: Why Alfalfa Is Not the Only Little Rascal for Bio-Agriculture Law. *Villanova Environmental Law Journal* 21: 383–407.

Stringam, G. R., and R. K. Downey. 1982. Effectiveness of Isolation Distances in Seed Production of Rapeseed (*Brassica napus*). *Agronomy Abstracts*: 136–137.

Suzuki, David. 2000. Experimenting with Life. http://www.yesmagazine.org/issues/food-for-life/356 (accessed January 6, 2012).

Swenarchuk, Michelle, and Canadian Environmental Law Association. 2003. *The Harvard Mouse and All That: Life Patents in Canada*: CELA Publication no. 454.

Takacs, David. 1996. *The Idea of Biodiversity: Philosophies of Paradise*. Baltimore: Johns Hopkins University Press.

Tanaka, Keiko, Arunas Juska, and Lawrence Busch. 1999. Globalization of Agricultural Production and Research: The Case of the Rapeseed Subsector. *Sociologia Ruralis* 39 (1): 54.

Tarrow, Sidney. 2005. *The New Transnational Activism*. Cambridge: Cambridge University Press.

Taverne, Dick. 2005. *The March of Unreason: Science, Democracy, and the New Fundamentalism*. Oxford: Oxford University Press.

Taylor, Arnold. 2010. Submission on Bill C-474 from Organic Farmer Arnold Taylor, addressed to Alex Atamanenko, MP for BC Southern Interior. http://oapf.saskorganic.com/pdf/Arnold_Taylor_C-474.pdf (accessed July 10, 2011).

Taylor, Michael R., and Jody S. Tick. 2001. *The StarLink Case: Issues for the Future*. Washington, DC: Resources for the Future/Pew Initiative on Food and Biotechnology.

Tesh, Sylvia Noble. 2000. *Uncertain Hazards: Environmental Activists and Scientific Proof*. Ithaca, NY: Cornell University Press.

Third World Network. 2002. Sustainability at the Crossroads: Which Way Forward? http://www.twnside.org.sg/title/twr145h.htm (accessed June 11, 2010).

Thompson, C. E., G. Squire, G. R. Mackay, J. E. Bradshaw, J. Crawford, and G. Ramsay. 1999. Regional Patterns of Gene Flow and Its Consequences for GM Oilseed Rape. In *Gene Flow and Agriculture: Relevance for Transgenic Crops*, ed. P. J. W. Lutman, 95–100. British Crop Protection Council Conference Proceedings 72. Farnham, Surrey, UK: British Crop Protection Council.

Tickell, Adam, and Jamie Peck. 2003. Making Global Rules: Globalisation or Neoliberalisation? In *Remaking the Global Economy: Economic-Geographical Perspectives*, ed. Jamie Peck and Henry Wai-Chung Yeung, 163–181. London: Sage.

Tilly, Charles. 1978. *From Mobilization to Revolution*. New York: McGraw-Hill Companies.

Touraine, Alain. 1981. *The Voice and the Eye: An Analysis of Social Movements*. Cambridge: Cambridge University Press.

Turner, R. Stephen. 2001. On Telling Regulatory Tales: rBST Comes to Canada. *Social Studies of Science* 31 (4): 475–506.

Turnley, Jessica Glicken. 2002. *Social, Cultural, Economic Impact Assessments: A Literature Review Prepared for the Office of Emergency and Remedial Response US Environmental Protection Agency*. http://www.epa.gov/superfund/community/involvement.htm (accessed January 6, 2012).

Turrent, Antonio, and Jose Antonio Serratos. 2004. Context and Background on Maize and Its Wild Relatives in Mexico. In *Maize and Biodiversity: Background Volume*. Montreal: Commission for Environmental Cooperation. http://www.cec.org/Page.asp?PageID=924&ContentID=2796 (accessed January 6, 2012).

UNEP Biosafety Working Group. 1997. Compilation of Government Submissions of Draft Text on Selected Items. UNEP/CBD/BSWG/3/3, August 15.

Unión de Científicos Comprometidos con la Sociedad. 2009. Letter Addressed to President of Mexico, Felipe de Jesús Calderón Hinojosa (English Translation). http://www.uccs.mx/doc/g/sciencetrmaize (accessed June 3, 2011).

Union of Concerned Scientists. 2002. Risks of Genetic Engineering.http://www.ucsusa.org/food_and_agriculture/science_and_impacts/impacts_genetic_engineering/risks-of-genetic-engineering.html (accessed May 18, 2011).

Union of Concerned Scientists. 2003. Genetic Engineering Techniques.http://www.ucsusa.org/food_and_agriculture/science_and_impacts/science/genetic-engineering-techniques.html (accessed May 18, 2011).

US Department of Agriculture. 2010. Glyphosate-Tolerant Alfalfa Events J101 and J163: Request for Nonregulated Status—Final Environmental Impact Statement. http://www.aphis.usda.gov/biotechnology/downloads/alfalfa/gt_alfalfa%20_feis.pdf (accessed January 6, 2012).

Van Acker, Rene C. 2004. Letter from Dr. Rene Van Acker to the Chief Justice of Canada regarding the Supreme Court of Canada's Ruling on the Case of *Schmeiser vs. Monsanto*. GRAIN. http://www.grain.org/research/contamination.cfm?id=172 (accessed May 31, 2010).

Vanclay, Frank. 2006. Principles for Social Impact Assessment: A Critical Comparison between the International and US Documents. *Environmental Impact Assessment Review* 26 (1): 3–14.

Vera Herrera, Ramón. 2004. En Defensa del Maíz (y el Futuro): Una Autogestión Invisible. Citizen Action in the Americas Series 13. Program of the Americas, Interhemispheric Resource Center. http://www.cipamericas.org/archives/1020 (accessed January 6, 2012).

Villar, Juan Lopez. 2001. Hands Off Our Seeds! Farmers' Rights Threatened by Biotech Industry. *Link* 97. http://www.foei.org/en/resources/link/97/e9717.html/?searchterm=percy%20schmeiser (accessed June 11, 2010).

Vilsack, Thomas J. 2010. Open Letter to Stakeholders. US Department of Agriculture.http://www.usda.gov/wps/portal/usda/usdahome?contentidonly=true&contentid=2010/12/0674.xml (accessed January 6, 2012).

Wainwright, Joel D., and Kristin L. Mercer. 2011. Transnational Transgenes: The Political Ecology of Maize in Mexico. In *Global Political Ecology*, ed. Richard Peet, Paul Robbins, and Michael J. Watts, 412–430. New York: Routledge.

Warick, Jason. 2001. GM Flax Seed Yanked Off Canadian Market—Rounded Up, Crushed. *StarPhoenix*, June 23. http://www.rense.com/general11/gm.htm (accessed July 6, 2011).

Warick, Jason. 2003. Lining Up against GM Wheat. *Saskatoon StarPhoenix*, August 9, E1.

Weber, Max. 1991. Science as a Vocation. In *From Max Weber: Essays in Sociology*, ed. Hans Heinrich Gerth and C. Wright Mills. New York: Oxford University Press. Originally published in 1919.

Willer, Helga, and Lukas Kilcher, eds. 2011. *The World of Organic Agriculture—Statistics and Emerging Trends 2011*. Bonn: International Federation of Organic Agriculture Movements.

Winickoff, David E., and Douglas M. Bushey. 2010. Science and Power in Global Food Regulation: The Rise of the Codex Alimentarius. *Science, Technology, and Human Values* 35 (3): 356.

Winickoff, David, Sheila Jasanoff, Lawrence Busch, and Robin Grove-White. 2005. Adjudicating the GM Food Wars: Science, Risk, and Democracy in World Trade Law. *Yale Journal of International Law* 30: 81.

Winner, Langdon. 1986. *The Whale and the Reactor: A Search for Limits in an Age of High Technology*. Chicago: University of Chicago Press.

Wirth, John D. 2003. Perspectives on the Joint Public Advisory Committee. In *Greening NAFTA: The North American Commission for Environmental Cooperation*, ed. David L. Markell and John H. Knox, 199–215. Stanford, CA: Stanford University Press.

Wise, Timothy A. 2007. *Policy Space for Mexican Maize: Protecting Agro-Biodiversity by Promoting Rural Livelihoods (Working Paper No. 07–01)*. Global Development and Environment Institute: Tufts University.

Woodhouse, Edward, David Hess, Steve Breyman, and Brian Martin. 2002. Science Studies and Activism: Possibilities and Problems for Reconstructivist Agendas. *Social Studies of Science* 32 (2): 297–319.

Wu, Felicia, and William P. Butz. 2004. *The Future of Genetically Modified Crops: Lessons from the Green Revolution*. Santa Monica, CA: RAND Corporation.

Wynne, Brian. 2007. Risky Delusions: Misunderstanding Science and Misperforming Publics in the GE Crops Issue. In *Genetically Engineered Crops: In-*

terim Policies, Uncertain Legislation, ed. Iain E. P. Taylor, 341–372. New York: Haworth Food and Agricultural Products Press.

Yashar, Deborah J. 1998. Contesting Citizenship: Indigenous Movements and Democracy in Latin America. *Comparative Politics* 31 (1): 23–42.

Yearley, Steven. 1992. Green Ambivalence about Science: Legal-Rational Authority and the Scientific Legitimation of a Social Movement. *British Journal of Sociology* 43 (4): 511–532.

Yoon, Carol Kaesuk. 2001. Genetic Modification Taints Corn in Mexico. *New York Times*, October 2, F7.

Ziff, Bruce. 2005. Travels with My Plant: *Monsanto v. Schmeiser* Revisited. University of Ottawa Law and Technology Journal 2 (2): 493–509.

Index